Results and Problems in Cell Differentiation

Series Editors:
W. Hennig, L. Nover, U. Scheer

29

Springer
Berlin
Heidelberg
New York
Barcelona
Hong Kong
London
Milan
Paris
Singapore
Tokyo

Siegfried Hekimi (Ed.)

The Molecular Genetics of Aging

With 53 Figures and 8 Tables

 Springer

Dr. Siegfried Hekimi
McGill University
Department of Biology
1205 Ave. Docteur Penfield
Montréal, Quebec H3A 1B1
Canada

ISSN 0080-1844
ISBN 3-540-66663-X Springer-Verlag Berlin Heidelberg New York

Library of Congress Cataloging-in-Publication Data.

The molecular genetics of aging/[edited by] Siegfried Hekimi. p. cm. – (Results and Problems in Cell Differentiation, ISSN 0080-1844; 29) Includes bibliographical references and index. ISBN 354066663X (hardcover) 1. Aging – Genetic aspects. I. Hekimi, Siegfried, 1956 – II. Series. QH607.R4 vol. 29 [QP86] 571.8′35 s – dc21 [571.8′78]

© Springer-Verlag Berlin Heidelberg 2000
Printed in Germany

Production: Angélique Gcouta
Cover Concept: Meta Design, Berlin
Cover Production: design & production, D-69121 Heidelberg
Typesetting: Scientific Publishing Services (P) Ltd, Madras
Printed on acid-free paper   SPIN: 10711213   39/3020GC 5 4 3 2 1 0

Preface

The molecular genetics of aging or life-span determination is an expanding field. One reason is because many people would consider it desirable if human life span could be extended. Indeed, it is difficult not to be fascinated by tales of the life and death of people who have succeeded in living a very long life. Because of this, we have placed at the head of this book the chapter by Perls et al. on *Centenerians and the Genetics of Longevity*. Perls and his coauthors convincingly argue that, while the average life expectancy might be mostly determined by environmental factors because the average person has an average genotype, extremely long life spans are genetically determined.

Of course, studying humans to uncover the genetics of aging is not ideal, not so much because one cannot easily perform experiments as because they live such a long time. This is why most of this book describes the current state of research with model organisms such as yeast, worms, flies, and mice. Jaswinski focuses on yeast and how metabolic activity and stress resistance affect the longevity of *Saccharomyces cerevisiae*. In the process, he discusses the concept of aging as applied to a unicellular organism such as yeast and the importance of metabolism and stress resistance for aging in all organisms.

In the following chapter, Sohal, Mockett, and Orr, whose experimental organism is *Drosophila*, give us a general perspective of the role of oxidative stress in aging. In this context, they argue that mutants whose extended life span is concomitant with a slow rate of aging do not provide true insight into the causes of aging because they do not alter the rate of aging with respect to the rates of other organismal parameters such as reproduction. In other words, mutations that make an organism live long by making it live slowly throw out the baby with the bath. In addition, they show how experiments with transgenic animals are subjected to many caveats. In particular, the presence of the transgene might induce a stress response or a slowing down of the rate of aging, that is independent of the biochemical activity of the protein encoded by the gene. Their clear analysis is a welcome reminder to all of us of how necessary it is to remain rigorous, in particular in a field that easily gets people excited.

The distinction between chronological age and developmental age is also addressed by Helfand and Rogina, who describe how gene expression during aging can be studied with great precision and resolution in *Drosophila*, particularly in the antennae.

Their work helps to address the question of whether short-lived mutants provide insight in the mechanisms of aging. While it is clear that mutations prolonging life span must alter a gene whose normal function limits life span, it is much less certain that each mutation that shortens life span accelerates the processes that normally underlie aging. Helfand and Rogina have identified and studied subtle biomarkers of aging, and their analysis, and the tools it provides, should help to assess whether a given mutation that shortens life span is associated with an increased rate of aging or whether it shortens life span through a novel pathology that is unrelated to the processes that occur during senescence in wild-type animals.

The next four chapters describe work on and insights into the aging process obtained through studies with the nematode *Caenorhabditis elegans*. In recent years, *C. elegans* has become an important model system for the study of aging, for two principal reasons, I believe. Firstly, worms live for only a short time, approximately 2 weeks on average, and their life and death is easy to score. This allows large-scale experiments to be performed and to be repeated several times within an acceptable time frame. Secondly, *C. elegans* is a self-fertilizing hermaphrodite, which means that comparisons can be made between strains that are fully isogenic except for the mutation under study. In the first worm chapter, I describe the action of three groups of genes: genes that affect developmental dormancy (the dauer larva pathway and *daf* genes), genes that affect the rate of living (*clk* genes), and genes that affect food intake (*eat* genes). Mutations in these three groups have also been studied in combination, in an effort to determine whether mutations that alter aging can be organized into genetic pathways in the manner that has been so successful with genes that affect development. I discuss these studies, their findings and their limits. Finally, I try to give a global view of how the genes under discussion might act together to determine life span by considering their action in the context of the organism as a whole.

One of the criticisms of using invertebrate model systems to study aging, as voiced by clinically inclined researchers, is that one cannot study relevant age-associated pathologies. The chapter by Herndon and Driscoll focuses on work in *C. elegans* toward understanding the genetic basis of apoptotic as well as necrotic cell death and what the relation of these processes to organismal aging might be. This might be particularly relevant given that some important age-associated diseases such as Alzheimer's disease are accompanied by loss of neuronal cell populations. Furthermore, recent results suggest that the molecular mechanisms of apoptosis are involved in the mechanisms underlying the development as well as the pathology of Alzheimer's. Herndon and Driscoll describe some of their findings with mutations that can suppress the increased prevalence of necrotic cell deaths produced by genetic manipulations. Strikingly, these suppressor mutations also increase organismal life span, suggesting that part of what limits life span is the increased prevalence of necrotic cell deaths. This resembles what we see with people where the medical alleviation of age-associated pathologies suffices to increase mean life span.

Possibly, one way forward is to explore whether a common mechanism or theme can link the findings with various mutants in worms and other organisms. Such thematic analyses are particularly important because of the difficulty in distinguishing which, among the variety of phenotypes displayed by most long-lived mutant strains, are causally linked to increased life span and which are accompanying phenomena. Lithgow explores the theme of stress resistance and Ishii and Hartman the related theme of the impact of oxidative stress on life span. These phenomena have traditionally been considered to be very important for life span and have therefore been studied in great detail by methods that did not involve the analysis of long-lived mutants. Testing the relevance of already well-documented phenomena by genetic means and the analysis of mutants is proving to be very powerful in this respect.

The last three chapters describe research using the mouse as model system. Although any finding with mice is likely to be the most relevant to the human condition, it is also true that mice are much larger, live much longer, and are much more expensive to maintain than yeast, worms, or flies. Because of this, it is much more difficult to use mice to carry out the sort of experimentation (mutant screens and aging experiments with large sample sizes and many different genotypes) that has been done successfully with the invertebrate model systems. However, mice provide tools and insights that invertebrates do not as yet, and might never, provide. These include the possibility of doing detailed biochemistry. Not only is it difficult to obtain a large amount of fresh material from whole animals, it is also very difficult in flies and virtually impossible in worms to obtain enough material from single organs. Most importantly, however, is the existence of a wealth of knowledge about the physiology and biochemistry of the animal and each of its organs and tissues. These features can be examined during aging and the biological and physiological consequences of mutations and treatments can thus be determined. All these advantages are well illustrated in these last three chapters. For example, Dollé, Giese, van Steeg, and Vijg describe the elegant and powerful techniques they have pioneered to examine genomic stability in aging. They address this question in two ways. First, they look at the accumulation of somatic mutations and its relative importance for the aging process. Second, they look into the link between mutagenesis and cancer, which may be the single most important age-related pathology.

All mutations known to date that prolong life span display a number of additional phenotypes, none perhaps more strikingly than the dwarf mice described by Bartke. It is very surprising to me that animals with such profound anatomical and physiological defects manage to live longer than their nonmutant counterparts. Bartke examines a number of physiological hypotheses to explain this pattern. For example, he points out that although individuals of larger species tend to live longer than those of smaller species (probably because of a difference in rate of living), the smaller individuals in a given species live longer than the larger individuals. This pattern is also seen among races of dogs. This notion seems to worry some people and I have even been interrogated by towering giants in scientific symposia about my opinion on the validity of this idea.

We finish the book with the chapter by Van Zant on the links between stem cells development and aging. Van Zant discusses what stem cells really are and describes their behavior during aging, also touching on experimental manipulations of telomeres and stem cells in vitro. He shows how the biology of stem cells might hold part of the key to what happens in an aging animal. These views are upheld by his identification of a genetic locus that appears to influence stem cell behavior and life span concurrently. It seems that in the scientific and lay public's mind, the research subject most clearly associated with the biology of aging is that of telomeres and stem cells. I suspect that the reasons for this lie in a conceptually easy idea and in the beauty of a word. The idea is that telomere shortening represents a counting mechanism, and the attraction is for the word "immortalized".

In a way we have come full circle. The book starts with a chapter on the basic phenomenon that some people live longer than others, probably because of their particular genetic constitution, and finishes with a chapter on a subject regarded by many as the best hope to overcome our genetic limitations.

Montréal, November 1999   *Siegfried Hekimi*

Contents

Centenarians and the Genetics of Longevity
Thomas Perls, Dellara F. Terry, Margery Silver, Maureen Shea,
Jennifer Bowen, Erin Joyce, Stephen B. Ridge, Ruth Fretts, Mark Daly,
Stephanie Brewster, Annibale Puca, and Louis Kunkel

Crossroads of Aging in the Nematode *Caenorhabditis elegans*
Siegfried Hekimi

Contributions of Cell Death to Aging in *C. elegans*
Laura A. Herndon and Monica Driscoll

Stress Response and Aging in *Caenorhabditis elegans*
Gordon J. Lithgow

Oxidative Stress and Aging in *Caenorhabditis elegans*
Naoaki Ishii and Philip S. Hartman

Mutation Accumulation In Vivo and the Importance of Genome Stability in Aging and Cancer
Martijn E. T. Dollé, Heidi Giese, Harry van Steeg, and Jan Vijg

Delayed Aging in Ames Dwarf Mice.
Relationships to Endocrine Function and Body Size
Andrzej Bartke

Stem Cells and Genetics in the Study of Development,
Aging, and Longevity
Gary Van Zant

Centenarians and the Genetics of Longevity

Thomas Perls[1], Dellara F. Terry[1], Margery Silver[1], Maureen Shea[1],
Jennifer Bowen[1], Erin Joyce[1], Stephen B. Ridge[2], Ruth Fretts[3],
Mark Daly[4], Stephanie Brewster[5], Annibale Puca[5], and Louis Kunkel[5]

1
Introduction

People have always been fascinated with extreme longevity. The ability to achieve extreme old age has been viewed as a miracle, as in the case of Methuselah, who lived 969 years, and as a curse in the case of Jonathan Swift's strudleburgs, in his book *Gulliver's Travels*. Claims of supercentenarians (age greater than 110 years) still capture the attention of the popular media and are accepted as valid without further investigation. Scientists and the lay public usually ask, "What's their trick?", expecting or perhaps hoping that there is a magic environmental factor that can be manipulated into a fountain of youth. Unsubstantiated and dangerous promises of eternal youth by antiaging practitioners who sell $20,000-per-year regimens of human growth hormone also profit from the belief held by many that immortality is a simple game of resetting the endocrinologic thermostat.

Feeding into this idea that aging can be manipulated environmentally is a highly popularized Danish study of monozygotic and dizygotic twins in which the heritability of life expectancy was only 20–30% (McGue et al. 1993; Herskind et al. 1996; Finch and Tanzi 1997; Rowe and Kahn 1998); but the oldest subjects in this study were in their mid to late 80s and the majority lived to average life expectancy (which in the United States is 76 years, about 74 years for men and 79 years for women). Therefore, differences in the environment accounted for 70% of the variability in age at death for those with an average life expectancy. If average humans are teleologically born

[1] Gerontology Division, Beth Israel Deaconess Medical Center, Harvard Division on Aging, 330 Brookline Ave, Boston, MA 02215, USA
[2] Department of Medicine, Columbia-Presbyterian Medical Center, New York, NY 10032, USA
[3] Department of Obstetrics and Gynecology, Beth Israel Deaconess Medical Center, Harvard Division on Aging, 330 Brookline Ave, Boston, MA 02215, USA
[4] Center for Genome Research at the Whitehead Institute for Biomedical Research, One Kendall Square, Building 300, Cambridge, MA 02139, USA
[5] Genetics Division, Howard Hughes Medical Institute, Children's Hospital and Harvard Medical School, 300 Longwood Avenue, Boston, MA 02115, USA

Results and Problems in Cell Differentiation, Vol. 29
Hekimi (Ed.): The Molecular Genetics of Aging
© Springer-Verlag Berlin Heidelberg 2000

with an average set of genetic polymorphisms, it will be differences in their habits and their environments that will explain the variability in their life expectancy. Unfortunately, these twin studies of average life expectancy say nothing of the ability to achieve extreme old age – that is to live another 20 years to age 100 and beyond. To do so requires a distinct genetic advantage, the arguments for which we outline below. Evidence has also been uncovered to show that families highly clustered for extreme longevity may be the key to discovering exactly what this distinct genetic advantage is.

2
Are Centenarians a New Phenomenon?

Prior to the 20th century, average life expectancy was about 45 years of age. However, one must distinguish between the definitions of average life expectancy and life span. Average life expectancy is the average age to which members of the population survive. Life span is the maximum age obtainable for the species and is defined by the age of the oldest living individual. In the case of humans, that individual was Madame Jeanne Calment, who died at the age of 122 years in August, 1997. Madame Calment had a tremendous responsibility; for every day she lived, she extended the human life span by a day.

Life span was probably not that different prior to the 20th century, although life expectancy was half of what it is today. There are numerous instances of people living well into their 90s reported as far back as ancient Greece. Tiziano Vecellio, known as Titian (1488–1576), lived to almost age 90. Leonardo da Vinci, born 1452, known for his scientific accuracy, drew several pictures of a 100-year-old man. Andrea della Robbia, a Florentine artist famous for his terra cottas, reportedly lived to age 90 (1435–1525) and around the same time, Michelangelo (1475–1564) lived to age 91. Hippocrates reportedly died in his mid-80s (460–377 B.C.), while Sophacles died in his mid-90s.

The question of whether centenarians are only a recent phenomena is important. If so, it would imply that environmental factor(s) are responsible. Genetic changes in the wild take place over the course of tens of thousands of years, not a few centuries. Given the historical record that centenarians, though much more rare in the past, are not a new phenomenon, this suggests a genetic etiology for longevity.

3
Centenarians Are the Fastest Growing Age Group

Among industrialized countries, the number of centenarians is increasing at an exceptionally rapid rate of about 8% per year. In comparison, population growth of 1% per year is the norm (Vaupel and Jeune 1995). In France, for instance, there were 200 centenarians in 1953, 3000 in 1989, and in 3 years, French demographers estimate there will be 6000 centenarians (Forette 1997). In 1994, Clarke reported that Queen Elizabeth had sent out birthday wishes to 300 centenarians in 1955, 1200 in 1970, and 3300 in 1987 (Clarke and Mittwoch 1994). At the turn of the century in the United States, approximately 1 per 100,000 was a centenarian and now it is about 1 per 10,000. This dramatic increase is probably the result of relatively recent public health measures, which have allowed people who would have otherwise succumbed to preventable or treatable causes of childhood or premature mortality to survive to much older age. At the turn of the century, one out of four children died from primarily infectious diseases. Mortality of both mother and child during childbirth was also significant. In addition, treatment of now readily reversible causes of death among older people is also making an important impact. As a result, many more people with genetic and environmental traits that facilitate survival to extreme old age are now able to achieve their potential life expectancy. Additionally, many more people in their eighties and nineties are living longer because of major improvements in medical care of older people.

One group that is not growing in number are the centenarians achieving purported ages of 120 years and beyond. In fact, with the exception of Jeanne Calment, they do not really exist. There are several geographical areas that have claimed inhabitants with extreme longevity, but after closer examination these claims have been found to be false. Vilacamba, in Ecuador, almost became a tourist attraction because natives claimed their water was a fountain of youth leading to the many super-centenarians in that region. Yogurt television commercials reported yogurt-eating centenarians living to 150 years and beyond in the Caucasus/Azerbaijan region. In the early 1980s American scientists traveled to these regions with high hopes and visions of discovering factors that slow down aging. Instead, after careful investigation, they found the ages to be off by a generation. The centenarians were either using birth or church certificates of aunts or uncles with the same names as their own or in other cases, there was no reasonable proof at all. Cases of extreme longevity require detailed scrutiny because they would be so incredibly rare. That is not to say that the elders of Azerbaijan are not worth studying. The potentially high prevalence of people reaching at least their 80s or even 90s in relatively good health despite third-world conditions is noteworthy.

4
Are Centenarians Different?

In 1825, Benjamin Gompertz proposed that mortality rate increased exponentially with age. Since then, most researchers accept that this rule indeed applies at younger adult ages for many species; however, at extreme old age an exception to the rule exists. At very old age, mortality begins to decelerate in species such as medflies (Carey et al. 1992), *Caenorhabdititis elegans* (Brooks et al. 1994), and humans (Thatcher et al. 1996). Survival calculations from a European database of 70 million people who reached at least age 80, and 200,000 who lived to at least 100 years of age were performed by Thatcher and colleagues. Their findings are illustrated in Fig. 1 (Thatcher et al. 1996). Line A depicts Gompertz's prediction of a constant exponential rise in mortality with age. Line B is the logistic curve that best fits the entire data set. These researchers attempted to find a mathematical function which would best predict line B. Line C is the logarithm of the mortality rate as a quadratic function of age fit to the data at ages 105 and higher. This equation accurately estimates the age-trajectory of human mortality up to about age 107 for females and age 105 for males. The quadratic curve fit to data at ages 105+ suggests a decline in mortality after age 110. Critics have argued that mortality deceleration may be an artifact of compositional change in heterogeneous populations. Studies of *Drosophila* data, however, demonstrate that a leveling off of death rates can occur when heterogeneity is minimized by rearing genetically homogeneous cohorts under very similar conditions (Vaupel et al. 1998).

Why does mortality decelerate at very old age? Most likely it is because frailer individuals drop out of the population, leaving behind a more robust

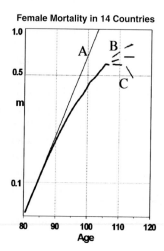

Fig. 1. Mortality rate associated with age. Line A is Gompertz's prediction. Line B depicts the deceleration in mortality observed at extreme old age

cohort which continues to survive. Because these frail individuals drop out of the population, the distribution of certain genotypes and other survival-related attributes in a cohort changes with older and older age. This selecting-out process is termed demographic selection.

The effect of demographic selection is exemplified by the drop out with extreme age of the apolipoprotein E ε-4 allele (Schächter et al. 1994). Rebeck and colleagues noted the frequency of the ε-4 allele to decrease markedly with advancing age (Rebeck et al. 1994). One of its counterparts, the ε-2 allele, becomes more frequent with advanced age. Presumably the drop out at earlier age of the ε-4 allele is because of its association with "premature" mortality secondary to Alzheimer's disease and heart disease. Recent investigation of the increased heritability of cognitive functional status at older age, also supports the possibility that genetic polymorphisms may play an increasing role with older age (McClearn et al. 1997).

A similar trend exists in the case of the apolipoprotein B locus, where investigators from the Italian centenarian study comparing 143 centenarians with younger controls demonstrated an association between specific multiallelic polymorphisms and extreme longevity (DeBenedictis et al. 1997). In another study, nonagenarian subjects had an extremely low frequency of HLA-DRw9 and an increased frequency of DR1. A high frequency of DRw9 and a low frequency of DR1 are associated with autoimmune or immune-deficiency diseases which can cause premature mortality (Takata et al. 1997). Recently, Tanaka and colleagues demonstrated single nucleotide substitutions in three mitochondrial genes that were present in the majority of a small centenarian sample but rare in the general population (Tanaka et al. 1997). This study requires verification in other populations given its small and homogeneous sample. Franceschi and colleagues, analyzing mtDNA haplogroups from 212 healthy centenarians and 275 younger controls, found that in male centenarians from Northern Italy, the J haplogroup was ten times more frequent than in controls (23% vs. 2%; $p = 0.005$) (pers. comm.).

Such findings suggest that centenarians would be ideal subjects for the discovery of other polymorphisms associated with a survival advantage. Furthermore, the apolipoprotein E ε-4 allele serves as an example of a polymorphism with an influence powerful enough to have a noticeable effect upon survival in the general population. Wachter recently addressed the question of how many other such polymorphisms might exist. He argued that, based upon the significant differences in mortality risk between people in their 90s versus those aged greater than 105, "there could be a relatively small number of genes – hundreds, not tens of thousands – which one has to not have in order to survive ad extrema". Of course, having the right polymorphisms may be as important as lacking the wrong ones (Wachter 1997).

5
The Centenarian Phenotype: Compressing Morbidity Towards the End of Life

In a population-based study of centenarians in New England, diseases normally associated with aging, such as Alzheimer's disease, have been noted to be at the least markedly delayed and sometimes absent in individuals of this cohort (Silver 1998). Functionally, centenarians as well as their children also appear to age relatively slowly. In the New England sample, 90% of centenarians were independently functioning at the age of 90 years (Perls et al. 1998a; Hitt et al. 1999). The underlying tenent is that centenarians, in order to achieve their age, must necessarily be healthy for the vast majority of their lives, and delay, if not escape, the diseases normally associated with aging which also cause premature mortality (Perls 1995).

Given preliminary findings that morbidity is relatively compressed in the centenarian cohort, centenarians should be ideal models of disease-free or disease-delayed aging. Despite this, relatively few studies of centenarians have been performed. Only a few have been population-based, thus negating the concern of selection bias. Most studies of centenarians have been specific in their scope, as in the study of dementia (Hauw et al. 1993; Powell 1994; Sobel et al. 1995), neuropathology (Hauw et al. 1986; Delaere et al. 1993; Bancher and Jellinger 1994; Fayet et al. 1994; Giannakopulos et al. 1995; Morrison and Hof 1997), thyroid function (Mariotti et al. 1993), body fat and metabolism (Paolisso et al. 1995), familiality and pedigree studies (Bocquet-Appel and Lucienne 1991; Perls et al. 1997, 1998; Alpert et al. 1998), cardiovascular and other risk factors (Barbagallo et al. 1995; Baggio et al. 1998), immune function (Mariotti et al. 1992; Effros et al. 1994; Francheschi et al. 1995), blood clotting (Mari et al. 1995; Mannucci et al. 1997), and pathological studies (Klatt and Meyer 1987). In terms of selection, some studies were intentionally biased towards the healthiest subjects (Beregi 1990; Beard 1991; Poon 1992), while in others, the rate of participation or recruitment methods raises the possibility of selection bias (Karasawa 1979; Allard 1991). Three studies have made good attempts at performing population-based centenarian studies in which enrollment success was high. Louhija, in his Ph.D. dissertation, had a remarkable recruitment success rate, locating and recruiting 185 of 190 centenarians in Finland (Louhija 1994). Perls and colleagues used 98% sensitive annual census records to locate and enroll 46 centenarians with an enrollment rate of 74%. The prevalence rate was 1 centenarian per 10,000 (Perls et al. 1999). Capurso and colleagues enrolled 382 randomly selected centenarians from a larger list of centenarians provided by the Italian census. The enrollment rate was 100% and the prevalence rate was 0.7 per 10,000 (Capurso et al. 1997). Thus far, no particular environmental trait, such as diet, economic status, or level of education has been found to significantly correlate with the ability to survive to *extreme* old age.

6
Evidence from Centenarians Supporting a Strong Genetic Influence upon Longevity

As noted throughout this book, tremendous advances are being made in understanding the genetics of aging and longevity. These findings based upon work in lower organisms are demonstrating that the manipulation of a few or even one gene(s) can lead to dramatic changes in an organism's rate of aging and life span. Despite the significant degree of homology between the human genome and the genomes of these lower organisms, many gerontologists believe that aging and suceptibility to diseases associated with aging in humans is the result of many complex interactions between hundreds if not thousands of genes, and therefore linkage or association studies in humans are impractical. Several discoveries which we have made regarding the highly selective nature of centenarians and the familiality of longevity have led us to the conclusion that such studies are indeed plausible.

7
Siblings of Centenarians Live Longer

In the course of recruiting centenarians in our population-based study, we frequently encountered subjects who reported similarly long-lived siblings. We therefore suspected that these siblings had genetic and environmental factors in common that conferred a survival advantage. To investigate the question of familiality of longevity further, we compared the longevity of siblings of 102 centenarians and siblings of a control group ($n = 77$) who were from a similar birth cohort born in 1896 but who died 27 years earlier at the age of 73 (Perls et al. 1998b). The siblings of the centenarians had a substantially greater chance of surviving to extremely old age compared to the siblings of the controls. This relative risk of survival steadily increased with age for siblings of the centenarians to the point that they had four times the probability of surviving to age 91 (Fig. 2). The relative risk for survival to older age continued to rise beyond age 91, though these larger differences were not statistically significant because of small numbers of siblings at these extreme ages. These findings indicate that there is a strong familial component to extreme longevity but they do not distinguish between the relative importance of shared environmental and genetic factors.

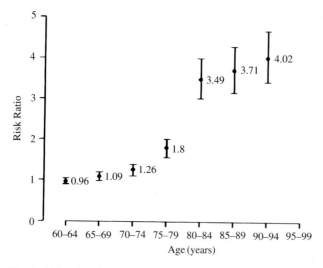

Fig. 2. Risk ratio of survival for siblings of centenarians versus siblings of 73-year olds

8
Parents of Centenarians also Achieve Unusually Old Age

For 106 centenarians in a population-based sample, the mean age of their mothers was 75.5 years ($n = 84$, SD $\pm$ 18.6 years) and for fathers it was 74.4 years ($n = 80$, SD $\pm$ 16.6 years). These averages are more than ten years higher than of the average life expectancy of the time, which for Massachusetts residents born in 1878–1882 who survived beyond the age of 20 years was 62 years for males and 63 years for females. A substantial proportion of parents achieved extreme old age; 21.6% of mothers and 13.5% of fathers survived to age $\geq$90 years.

9
Four Families with Clustering for Extreme Longevity

In gathering pairs of siblings as part of our Centenarian Study, we identified four families that clearly demonstrate segregation for extreme old age. We set out to expand these pedigrees and to determine if the clustering could be attributed to chance or if genetics must be playing a causative role. The pedigrees demonstrating vertical transmission of extreme longevity are shown in Fig. 3. Omitted family members died at age 18 years or younger, or died at an age less than 90 years because of trauma. The illustrated gender of certain members was altered for anonymity. Ages were validated using vital

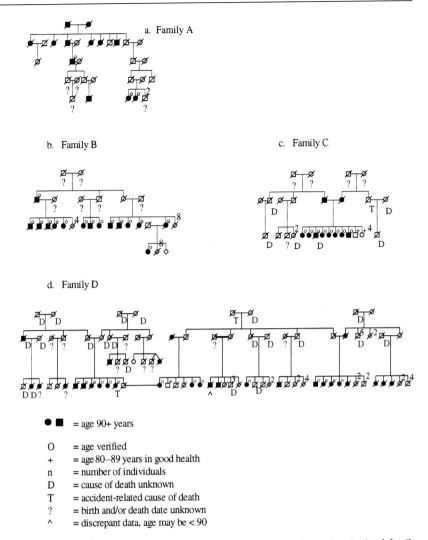

● ■ = age 90+ years

O = age verified
+ = age 80–89 years in good health
n = number of individuals
D = cause of death unknown
T = accident-related cause of death
? = birth and/or death date unknown
^ = discrepant data, age may be < 90

Fig. 3. Four families demonstrating vertical transmission of extreme longevity. Omitted family members died at age 18 years or younger, or died at an age less than 90 years because of accidental trauma. The illustrated gender of certain members was altered for anonymity. Ages were validated using vital records and U.S. Federal census entries. **a,** Family A is composed of 1 male and 4 females aged 100 or older in one generation living in the 17th and 18th centuries. **b,** In family B, the individuals of note were born in the 19th or early 20th century and 7 are centenarians. **c,** In family C, there is a sibship of 13 children with 8 reaching extreme old age (range: 90 to 102 years old). **d,** In family D, there are two branches linked together by a marriage in the 3rd generation. These different branches originated from the same small region in Norway. In the 3rd generation, 23 of 46 individuals achieved extreme old age (range: 90 to 106 years old)

records and US Federal census entries. Family A is composed of one male and four females aged 100 or older in one generation living in the 17th and 18th centuries. In family B, the individuals of note were born in the 19th and 20th centuries and seven are centenarians. In family C, there is a sibship of 13 children with 8 reaching extreme old age (range: 90 to 102 years old). In family D, there are two branches linked together by a marriage in the 3rd generation. In the 3rd generation, 23 of 46 individuals achieved extreme old age (range: 90 to 106 years old).

9.1
Mathematical Analysis

Cohort life tables for the years 1900, 1850 and 1801 were used to estimate the probability of individuals in Fig. 3 surviving to their specified ages. For earlier birth cohorts, such as those encountered in the family A pedigree, the 1801 cohort life table was used as a conservative estimate of probability. These specific probabilities were then used to calculate a binomial probability of obtaining N individuals achieving their specified ages from a random sample of M individuals belonging to specific birth cohorts. Probabilities were calculated for the single most impressive generation of each family. Probabilities would be even lower if the individuals achieving extreme old age from other generations were also taken into account.

The random chance of encountering the six siblings age 90 and older in family A is one in 10^9. In family B, there are three sibships that compose all grandchildren of two individuals. The chance of 13 of the 20 grandchildren living past 90 is about one in 10^{18}. In one of the three sibships, 5 of 16 siblings achieved age 100 or older, also an extremely rare encounter if left to chance. In the case of family C, the chance of 8 siblings reaching at least 90 years is less than one in 10^{13}. Dominant inheritance is a possibility in this pedigree, given that both parents are affected and the children are largely or entirely affected. Though pedigree D appears to represent three unrelated branches, these branches originate from the same small region of Norway and therefore there may be a common ancestor. Nonetheless, treating the branches separately, the chance of encountering the 9 of 14 grandchildren reaching at least age 90 in the left branch is one in 10^{12}. The middle branch's grandparents had 12 of 38 grandchildren live to at least age 90, a chance occurrence of one in 10^{11}. The right branch's grandparents had 8 of 26 also living to at least age 90; a chance observation of one in 10^8.

The above probability values are smaller than 1 per the number-of-families-in-the-world today, so clearly there is familial aggregation that cannot be explained by random chance. Several points argue in favor of shared genetic factors rather than environmental factors, effecting such a survival advantage. Two of the four families include cousins achieving extreme old age and these relatives are unlikely to have a common childhood environment. Also, the four discribed families come from distinct backgrounds. While this implies genetic as well as environmental diversity, one really cannot imagine any environmental components shared by these families that would be responsible for extreme longevity. One would not necessarily expect that the families had the same genetic cause, but only that genetics plays an important role.

10
Middle Aged Mothers Live Longer:
An Evolutionary Link Between Reproductive Success and Longevity-Enabling Genes

As we reviewed the pedigrees of a number of our centenarian subjects living in the suburban Boston area, we came across a substantial number of women who had children in their 40s. There was even one that had a child at the age of 53 years. This struck us as unusual, given that maternal age greater than 40 is a relatively rare event. Less than 3% of births occur in women 40 years of age or older (Fretts et al. 1995). In 1995 the birth rate of American women 40–44 years was 6.6 per 1000 women, and 0.3/1000 for women 45–49 (Ventura et al. 1997). However, a history of older maternal age among our centenarian subjects made sense to us since aging relatively slowly is a likely necessary characteristic of achieving extreme age, and women who do so, should be able to bear children at an older age.

We went on to compare 78 female centenarians with a similar birth cohort of 54 women born in 1896, but who died at the age of 73 years in 1969. By collecting data on a similar birth cohort, we were able to minimize concerns about temporally related influences upon fertility such as health and contraception-related trends. We found that 19.2% of the centenarians had children at age 40 years or older compared to 5.5% of the women who lived to age 73 (Fig. 4). We concluded that if you are a woman who naturally had a child in her 40s, you are four times more likely to live to a 100 years old rather than to die at the age of 73 (Perls et al. 1998b). However, we believe that it is not the act of having a child in your 40s that promotes long life, but rather it is an indicator that the woman's reproductive system is aging slowly. A slow rate of aging would therefore bode well for the woman's subsequent ability to achieve very old age.

What are the factors that link the slowly aging reproductive system and the ability to reach extreme age? Ideally, to identify the reproductive factors that are associated with longevity we would like to know the status of various reproductive factors such as age at menarche, cycle regularity, number of spontaneous abortions, and age of menopause. Unfortunately, obtaining this

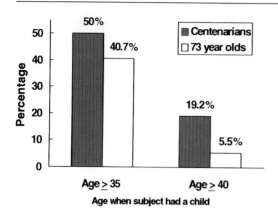

Fig. 4. Frequency of centenarian women and women surviving to age 73 years who gave birth at age 35 or older and at age 40 or older

type of information from relatives of the deceased is difficult and unreliable. During the first quarter of this century, fertility-enhancing interventions for older women were not available. Under these circumstances, knowing when a woman last had a child is our best estimate of her premenopausal status and therefore reflects her natural ability to conceive later in life. Relatively delayed menopause, like pregnancy after age 40, may be a marker for aging slowly and the subsequent ability to achieve extreme longevity. This finding is interesting not only for its potential value in predicting individuals predisposed to extreme longevity, but also because it has implications regarding the theoretical basis of menopause and human life span.

The age of menopause was noted to be linked to longevity by Snowdon et al., who studied 5287 White female Seventh-Day Adventists (Snowden et al. 1989). The women who experienced menopause prior to the age of 40 had nearly twice the risk of dying during the study period compared to women who had experienced natural menopause at ages 50–54. This was true even after correcting for smoking, weight, reproductive history, and estrogen replacement. Because a history of estrogen replacement did not change these findings, it is possible that prolonged endogenous estrogen exposure is not the only explanation for the increased longevity of women with relatively delayed menopause.

10.1
What Determines When a Woman Will Go Through Menopause?

Given available records, we know in Western societies the median age of menopause has remained relatively constant over at least the past 100 years and occurs on average, at about age 51 (Napier 1987). In contrast to this stability, the age of menarche has declined significantly as health and nutrition has improved. The age of menopause is influenced by both environmental and inherited factors. Smoking, the most common environmental toxin to ovarian function, causes earlier menopause as well as premature mortality (McKinlay et al. 1985). Chemotherapy, radiation, and surgery are less common causes of premature ovarian failure. In terms of inheritance, the influence of genetic factors is just becoming elucidated. A recent study of mothers and daughters found that the age of menopause in the mother was a significant predictor of the age of menopause in their daughters (Torgerson et al. 1997). This is particularly true of early menopause. There appear to be genes on the X-chromosome that are required for maintaining ovarian function. Deletions of important genetic information found on this chromosome have been associated with various degrees of premature ovarian failure and have been identified in mother–daughter pairs (Veneman et al. 1991; Tharapel et al. 1993). The genes that regulate when a woman goes through menopause may be closely linked to genes that regulate how fast we age.

10.2
Menopause: An Adaptive Response

Menopause may act evolutionarily as a sentry to protect the aging woman from the hazards of childbirth. We support Doris and George Williams' theory, developed in 1956, that as humans evolved and became able to achieve older and older ages, there came a point when survival during childbirth began to decline as a function of further aging and increased frailty (Williams and Williams 1957). Females, who by virtue of some genetic mutations became infertile prior to the age of marked mortality risk, had a survival advantage over those females that did not have this series of mutations. An equally important advantage of continued survival would be that the mother could continue to raise and assure the survival of her children beyond puberty as well as to assist in the care of her children's children. From an energy allocation point of view, there probably comes a point with older age (and decreasing energy) when it becomes more efficient to care for the children one already has and to perform other work in the society than to devote that energy to pregnancy, childbirth, and breast feeding. In primitive hunter-gatherer societies, postreproductive women perform a large portion of the work. Therefore, menopause provides a survival advantage and a means of better assuring the passing down of one's genes to subsequent generations. Williams and Williams thus called menopause an "adaptive response" to the increased mortality risk associated with childbirth.

10.3
Why Menopause Does Not Occur in Other Mammals?

In other mammals, giving birth is literally more straightforward and any associated mortality risk is relatively low even at advanced age. In humans, on the other hand, the birth canal has several twists and turns that developed coincidentally with the evolution of erect posture. It is this tortuous and cumbersome birth canal in humans that causes much of the mortality risk associated with childbirth. Because there is relatively little mortality risk associated with bearing young in other mammals, there is no selective advantage for the development of infertility (menopause) to assure the mother's survival and the survival of the young she has already produced.

The pilot whale, which also spends a significant proportion of time in a postreproductive state, is one exception to the rule (Austad 1994). This example of convergent evolution (when two unrelated species have a similar characteristic) may be due to chance, however it is interesting that the pilot whale also spends a significant amount of energy rearing its young. The mother pilot whale will suckle its young for up to 14 years after birth. In this setting, menopause may make sense where raising the young requires a significant period of time and energy before they are independent of their parent.

10.4
Nonhuman Data Supporting the Association
Between Delayed Reproductive Senescence and Increased Longevity

Our observation in humans that fecundity at older age is associated with longer life expectancy correlates with those made by Michael Rose in his selection experiments of fruit flies in which the ability to produce eggs later in life also correlates with greater life expectancy (Hutchinson and Rose 1991). Working with millions of flies, Dr. Rose and his colleagues selected out and bred flies that were born from eggs laid by the oldest females. With each subsequent generation, this selection process yielded older and older flies and the life span was increased.

10.5
Alternative Explanations for Why Menopause Occurs

Not all evolutionary biologists agree that menopause evolved because it provided a selective advantage. Since the average life expectancy has increased considerably within the last 100 years, some believe that menopause is a relatively recent phenomenon. These biologists assert that humans have "outlived" their ovaries and thus menopause is simply an artifact of an unexpected recent increase in life span beyond reproductive age. Firstly, though average life expectancy has increased markedly in just the past century, the human life span has been significantly longer than the age of menopause, probably since the time menopause evolved. There is no evidence to indicate that we have done something special as a species in even the past millennium that would facilitate a doubling or tripling of the human life span. Certainly there is evidence from Ancient Greece indicating elder statesmen living well into their 70s and early 80s. Secondly, if menopause was simply an artifact, we would not expect, over the course of evolution, a natural selection for genes that influence when menopause occurs. Contrary to this supposition, genetics does appear to play a role in when menopause occurs; and finally, the nonadaptive hypothesis begs the question of why would the reproductive system fail long before other systems, such as the cardiovascular system?

Alan Rogers, studying data from agrarian Taiwan collected in 1906, generated a mathematical model to estimate the genetic contribution of women with and without menopause (Rogers 1993). He estimated more than one out of ten women would have to die from childbirth in order for there to be a net decrease in a woman's genetic contribution to subsequent generations. If maternal mortality was less than 10% then he supposed that menopause would not be particularly adaptive. However, Rogers did not estimate and include in his mathematical model a variable that would sufficiently reflect the maternal energy that is required to ensure survival for her offspring and her children's offspring. Any estimate of energy required to raise a child to independence is likely to be underestimated. We argue that if both the maternal mortality and maternal energy are considered the evolution of menopause would be adaptive and increase the success rate for genetic contribution to future generations.

10.6
Why Is the Human Life Span 122 Years
and What Is the Evolutionary Advantage for Living to Such an Age?

Based upon our observations, the slower a woman ages, the longer period of time and greater opportunity she has to produce children and thus contribute her genes to the gene pool. Despite the fact that menopause must still occur, a continued slow rate of aging and perhaps also a decreased susceptibility to diseases associated with aging that can cause maternal mortality would therefore allow a woman to achieve extreme longevity. From a Darwinian point of view, we assert that there is no selective advantage for humans to have a life span of approximately 100 years. Rather, attaining such very old age is simply a by-product of the genetic forces that maximize the length of time during which women can bear children. Just because a woman goes through menopause does not mean that the genes which allowed her to age slowly and have a decreased predisposition to age-related diseases suddenly turn off; these genes continue to exert their influence, and thus enable her to achieve extremely old age. In other words, menopause may be the evolutionary fulcrum that determines human life span. Where does that leave men in the scheme of things? We (Perls and Fretts) would argue that the purpose of males in our species is simply to pass down genes to their female offspring that facilitate the woman to age as slowly as possible.

10.7
What If We Removed the Selective Force for Maximizing Life Span?

Due to the tremendous advances of 20th century obstetrics, mortality risk for the mother during childbirth has markedly declined. With this decline in risk, there is no longer the pressure to age slowly in order to maximize the period of time during which women have the opportunity to bear children. Therefore, it is unlikely that natural selection forces will promote further expansion of the human life span.

10.8
The Association Between Longevity-Enabling Genes and Genes Which Regulate Reproductive Health

The close relationship between genes that regulate reproductive fitness in women and genes that regulate rates of aging and susceptibility to diseases associated with aging is intriguing. To date, there have been no specific genes identified which promote later menopause; presumably they exist and are responsible for prolonged ovarian function. These ovarian function-promoting genes may be closely associated with longevity-enabling genes or they may somehow be one and the same. Perhaps such genes will be identified as the result of linkage and positional cloning studies of centenarian sib-pairs and families with multiple centenarians.

11
In Our Near Future

The evaluation of known candidate genes for polymorphisms occurring at frequencies which differ between centenarians and ethnically matched controls is another important approach, as illustrated by the afore mentioned work with apolipoprotein E and its major polymorphisms. Numerous researchers have their favorite candidate genes for investigating the modulators of aging and its associated diseases (Schächter 1998). Clearly, the discovery of candidate genes in lower organisms is crucial for the identification of genes to be studied in human cohorts such as centenarians. The explosion of findings currently occurring from these different approaches appears to indicate that we should have key answers to the molecular genetics of aging puzzle in the very near future.

Acknowledgments. We are indebted to the following: The Alzheimer's Association Darrell and Jane Phillippi Faculty Scholar Award, the National Institute on Aging (AG-00294 and R0-3), The Neurosciences Education and Research Foundation, and The Paul Beeson Faculty Scholar in Aging Research Award.

References

Allard M (1991) A la recherche du secret des centeraires. Le Cherche-Midi, Paris

Alpert L, DesJardines B, Vaupel J, Perls T (1998) Extreme longevity in two families. A report of multiple centenarians within single generations. In: Jeune B, Vaupel J (eds) Age validation of the extreme old. Odense monographs on population aging 4, Odense University Press, Odense

Austad SN (1994) Menopause, an evolutionary perspective. Exp Gerontol 29:255–263

Baggio G, Donazzan S, Monti D, Mari D, Martini S, Gabelli C, Dalla Vestra M, Previato L, Guido M, Pigozzo S, Cortella I, Crepaldi G, Franceschi C (1998) Lipoprotein(a) and lipoprotein profile in healthy centenarians: a reappraisal of vascular risk factors. FASEB J 12:433–437

Bancher C, Jellinger KA (1994) Neurofibrillary tangle predominant form of senile dementia of Alzheimer type: a rare subtype in very old subjects. Acta Neuropathol 88:565–570

Barbagallo CM, Averna MR, Frada G, Chessari S, Mangiacavallo G, Notarbartolo A (1995) Plasma lipid apolipoprotein and Lp(a) levels in elderly normolipidemic women: relationships with coronary heart disease and longevity. Gerontology 41:260–266

Bocquet-Appel JP, Lucienne J (1991) La transmission familiale de la longévité à Arthez d'Asson (1686–1899). Population 2:327–347

Brooks A, Lithgow GJ, Johnson TE (1994) Mortality rates in a genetically heterogeneous population of *Caenorhabditis elegans*. Science 263:668–671

Beard BB (1991) Centenarians, the new generation. Greenwood Press, New York

Beregi E (1990) Centenarians in Hungary. A social and demographic study. Interdisciplinary topics in gerontology. 27. Karger, Basel

Capurso A, D'Amelio A, Resta F, et al. (1997) Epidemiological and socioeconomic aspects of Italian centenarians. Arch Gerontol Geriatr 25:149–157

Carey JR, Liedo P, Orzoco D, Vaupel JW (1992) Slowing of mortality rates at older ages in large medfly cohorts. Science 258:457–461

Clarke CA, Mittwoch U (1994) Puzzles in longevity. Perspect Biol Med 37:327–336

De Benedictis G, Falcone E, Rose G, Ruffolo R, Spadafora P, Baggio G, Bertolini S, Mari D, Mattace R, Monti D, Morellini M, Sansoni P, Franceschi C (1997) DNA multiallelic systems reveal gene/longevity associations not detected by diallelic systems. The APOB locus. Hum Genet 99:312–318

Delaere P, He Y, Fayet G, Duykaerts C, Hauw JJ (1993) A-4 deposits are constant in the brain of the oldest old: an immunocytochemical study of 20 French centenarians. Neurobiol Aging 14:191–194

Effros RB, Boucher N, Porter V, Zhu XM, Spaulding C, Walford RL, Kronenberg M, Cohen D, Schächter F (1994) Decline in CD28(+) T cells in centenarians and in long term T cell cultures: a possible cause for both in vivo and in vitro immunosenescence. Exp Gerontol 29:601–609

Fayet G, Hauw JJ, Delaere P, He Y, Duykaerts C, Beck H, Forette F, Gallinari C, Laurent M, Moulias R, Piette F, Sachet A (1994) Neuropathologie de 20 centenaires. I Donnés cliniques. Rev Neurol 150 : 16–21

Finch CE, Tanzi RE (1997) Genetics of aging. Science 278 : 407–411

Forette B (1997) Centenarians: health and frailty. In: Robine JM, Vaupel JW, Jeune B, Allard M (eds) Longevity: to the limits and beyond. Springer, Berlin Heidelberg New York

Franceschi C, Monti D, Sansoni P, Cossarizza A (1995) The immunology of exceptional individuals: the lesson of centenarians. Immunol Today 16 : 12–16

Fretts RC, Schmittdiel J, McLean FH, Usher RH, Goldman MB (1995) Increased maternal age and the risk of fetal death. N Engl J Med 333 : 953–957

Giannakopulos P, Hof PR, Vallet PG, Giannakopoulos AS, Charnay, Bouras C (1995) Quantitative analysis of neuropathologic changes in the cerebral cortex of centenarians. Progr Neuro-Psychopharmacol Biol Psychiat 19 : 577–592

Hauw JJ, Vignolo P, Duykaerts C, Beck M, Forette F, Henry JF, Laurent M, Piette F, Sachet A, Bethaux P (1986) Etude neuropathologique de 12 centenaires: la fréquence de la démence sénile de type Alzheimer n'est pas particulièrement élevée dans ce groupe de personnes tres âgées. Rev Neurol 142 : 107–115

Hauw JJ, Delacre P, Fayet G, He Y, Costa C, Seilhean D, Duykaerts C, Beck H, Forette F, Gallinari C, Laurent M, Moulias R, Piette F, Sachet A (1993) The centenarian's brain. Sandoz 17–25

Herskind AM, McGue M, Holm NV, Sorensen TI, Harvald B, Vaupel JW (1996) The heritability of human longevity: a population-based study of 2872 Danish twin pairs born 1870–1900. Hum Genet 97 : 319–323

Hitt R, Young-Xu Y, Perls T (1999) Centenarians: the older you get, the healthier you've been. Lancet (In press)

Hutchinson EW, Rose MR (1991) Quantitative genetics of postponed aging in *Drosophila melanogaster*. I. Analysis of outbred populations. Genetics 127 : 719–727

Karasawa A (1979) Mental aging and its medico-social background in the very old Japanese. J Gerontol 34 : 680–686

Klatt EC, Meyer PR (1987) Geriatric autopsy pathology in centenarians. Arch Pathol Lab Med 111 : 367–369

Louhija J (1994) Finnish Centenarians. Academic dissertation. University of Helsinki, Helsinki

Mannucci PM, Mari D, Merati G, Peyvandi F, Tagliabue L, Sacchi E, Taioli E, Sansoni P, Bertolini S, Franceschi C (1997) Gene polymorphisms predicting high plasma levels of coagulation and fibrinolysis proteins: a study in centenarians. Arterioscl Thromb Vasc Biol 17 : 755–759

Mari D, Mannucci PM, Coppola R, Bottasso B, Bauer KA, Rosenberg RD (1995) Hypercoagulability in centenarians: the paradox of successful aging. Blood 85 : 3144–3149

Mariotti S, Sansoni P, Barbesino G, Caturegli P, Monti D, Cossarizza A, Giacomelli T, Passeri G, Fagiolo U, Pinchera A, Franceschi C (1992) Thyroid and other organ-specific autoantibodies in healthy centenarians. Lancet 339 : 1506–1508

Mariotti S, Barbesino G, Caturegli P, Bartalena L, Sansoni P, Fagnoni F, Monti D, Fagiolo U, Franceshi C, Pinchera A (1993) Complex alteration of thyroid function in healthy centenarians. J Clin Endocrinol Metab 77 : 1130–1134

McClearn GE, Johansson B, Berg S, Pedersen NL, Ahern F, Petrill SA, Plomin R (1997) Substantial genetic influence on cognitive abilities in twins 80 or more years old. Science 276 : 1560–1563

McGue M, Vaupel JW, Holm N, Harvald B (1993) Longevity is moderately heritable in a sample of Danish twins born 1870–1880. J Gerontol Biol Sci 48 : B237–B244

McKinlay SM, Bifano NL, McKinlay JB (1985) Smoking and age at menopause. Ann Intern Med 103 : 350–356

Morrison JH, Hof PR (1997) Life and death of neurons in the aging brain. Science 278 : 412–419

Napier ADL (1987) The Menopause and its Disorders. Scientific Press, London

Paolisso G, Gambardella A, Balbi V, Ammendola S, Damore A, Varricchio M (1995) Body composition, body fat distribution and resting metabolic rate in healthy centenarians. Am J Clin Nutrit 62:746–750

Perls TT (1995) The Oldest Old. Sci Am 272:70–75

Perls T, Alpert L, Fretts R (1997) Middle aged mothers live longer. Nature 389:133

Perls TT, Bochen K, Freeman M, Alpert L, Silver MH (1998) The New England Centenarian Study: validity of reported age and prevalence of centenarians in an eight town sample. In: Jeune B, Vaupel J (eds) Age validation of the extreme old. Odense Monographs on Population Aging 4. Odense University Press, Odense

Perls T, Wager C, Bubrick E, Vijg J, Kruglyak L (1998) Siblings of centenarians live longer. Lancet 351:1560

Perls TT, Bochen K, Freeman M, Alpert L, Silver MH (1999) Validity of reported age and prevalence of centenarians in a New England sample. Age and Ageing 28:193–197

Poon LW (1992) The Georgian Centenarian Study. Baywood, Amytyville, New York

Powell AL (1994) Senile dementia of extreme aging – a common disorder of centenarians. Dementia 5:106–109

Rebeck GW, Perls TT, West HL, Sodhi P, Lipsitz LA, Hyman BT (1994) Reduced apolipoprotein epsilon 4 allele frequency in the oldest old. Alzheimer's patients and cognitively normal individuals. Neurology 44(8):1513–1516

Rogers AR (1993) Evol Ecol 7:406–420

Rowe JW, Kahn RL (1998) Successful aging. New York: Random House

Schächter F, Faure-Delanef L, Guenot F, Rouger H, Froguel P, Lesueur-Ginot L, Cohen D (1994) Genetic associations with human longevity at the APOE and ACE loci. Nat Genet 6:29–32

Schäcter F (1998) Causes, effects and constraints in the genetics of human longevity. Am J Hum Genet 62:1008–1014

Silver M, Newell K, Growdon J, Hyman BT, Hedley-Whyte ET, Perls T (1998) Unraveling the mystery of cognitive changes in extreme old age: correlation of neuropsychological evaluation with neuropathological findings in centenarians. International Psychogeriatr 10:25–41

Snowden DA, Kane RL, Beeson WL (1989) Is early natural menopause a biological marker of health and aging? Am J Pub Health 79:709–714

Sobel E, Louhija J, Sulkava R, Davanipour Z, Kontula K, Miettinen H, Tikkanen M, Kainulenen K, Tilvis R (1995) Lack of association of apolipoprotein E allele epsilon 4 with late-onset Alzheimer's disease among Finnish centenarians. Neurology 45:903–907

Takata H, Suzuki M, Ishii T, Sekiguchi S, Iri H (1997) Influence of major histocompatibility complex region genes on human longevity among Okinawan-Japanese centenarians and nonagenarians. Lancet 2:824–826

Tanaka M, Gong JS, Zhang J, Yoneda M, Yagi K (1997) Mitochondrial genotype associated with longevity. Lancet 351(9097):research letter

Tharapel AT, Anderson KP, Simpson (1993) Deletion (X) (q26.2-q28) in a proband and her mother: molecular characterization and phenotypic–karyotypic deductions. Am J Hum Genet 52:463–471

Thatcher AR, Kannisto V, Vaupel JW, Yashin AI (1996) The force of mortality from age 80 to 120. University Press, Odense, Denmark

Torgerson DJ, Thomas RE, Reid DM (1997) Mothers and daughters menopausal age: is there a link? Eur J Obstet Gyn 74:63–66

Vaupel JW, Jeune B (1995) The emergence and proliferation of centenarians. In: Jeune B, Vaupel JW (eds) Exceptional longevity: From prehistory to the present. pp 109–116. Springer, Berlin Heidelberg New York

Vaupel JW, Carey JR, Christensen K, Johnson TE, Yashin AI, Holm NV, Iachine IA, Kannisto V, Khazaeli AA, Liedo P, Longo VD, Zeng Y, Manton KG, Curtsinger JW (1998) Biodemographic trajectories of longevity. Science 280:855–860

Veneman TF, Beverstock GC, Exalto N, Mollevanger P (1991) Premature menopause because of an inherited deletion in the long arm of the X-chromosome. Fertil Steril 55:631–633

Ventura JS, Martin JA, Curtin SC (1997) Report of final natality statistics, 1995. Centers for Disease Control and Prevention/National Center for Health Statistics. Bethesda

Wachter KW (1997) Between Zeus and the salmon: Introduction. In: Wachter KW, Finch CE (eds) Between Zeus and the salmon. The biodemography of longevity. National Academy Press, Washington, DC

Williams GC, Williams DC (1957) Natural selection of individually harmful social adaptations among sibs with special reference to social insects. Evolution 11:32–39

Coordination of Metabolic Activity
and Stress Resistance in Yeast Longevity

S. Michal Jazwinski

Summary: The genetic analysis of longevity in yeast has revealed the importance of metabolic control and resistance to stress in aging. It has also shown that these two physiological processes are interwoven. Molecular mechanisms underlying the longevity effects of metabolic control and stress resistance, as well as genetic stability, are emerging. The yeast *RAS* genes play a substantial role in coordinating at least the first two of these processes. Numerous correlates can be found between the physiological processes involved in yeast aging and aging in *Caenorhabditis elegans* and in *Drosophila*, and the dietary restriction paradigm in mammals.

1
Introduction

The yeast *Saccharomyces cerevisiae* has been a favorite model of cell and molecular biologists for the elucidation of fundamental biological processes for more than two decades. An important ingredient of this popularity is the facility of genetic analysis in this unicellular eukaryote. It is therefore not surprising that this organism has gained respect as a tool for the dissection of the molecular mechanisms of aging during the past 10 years. The finite life span of individual yeast cells was first adequately described by Mortimer and Johnston (1959), but the studies of Barton (1950) presaged this. A coherent cataloging of the phenomenology of yeast aging had to wait another 40 years (Jazwinski 1990a).

It would appear that the aging process differs substantially between yeast and mammals. On the other hand, a comparison of yeast cells and mammalian cells senescing in culture yields several similarities (Jazwinski 1990b), prompting the suggestion that yeasts are a model for the senescence of individual cells of a metazoan. These similarities are superficial. Mammalian

Department of Biochemistry and Molecular Biology, Louisiana State University Medical Center, New Orleans, LA 70112, USA

Results and Problems in Cell Differentiation, Vol. 29
Hekimi (Ed.): The Molecular Genetics of Aging
© Springer-Verlag Berlin Heidelberg 2000

cells can undergo a characteristic number of rounds of population doublings, after which they cease division, prompting the term replicative sensescence (Smith and Pereira-Smith 1996). Individual lineages of cells actually withdraw from the cell cycle; thus, this phenomenon is also referred to as clonal senescence. Individual yeast cells divide a limited number of times (Mortimer and Johnston 1959). However, their progeny have the capacity for a full replicative life span, so that the population or culture is immortal, as are individual yeast clones.

The differences between yeast and mammalian cells do not stop there. The telomeres of human chromosomes shorten as cells undergo successive population doublings (Harley et al. 1990). This has been shown recently to present a limit to their replicative capacity (Feng et al. 1995; Bodnar et al. 1998). Yeast telomeres do not shorten with replicative age (D'mello and Jazwinski 1991). Furthermore, artificial lengthening of telomeres in yeast curtails their replicative life span (Austriaco and Guarente 1997).

The distinction between yeast cells and cells from multicellular organisms becomes understandable when one considers the fact that the yeast cell is the organism. Thus, telomere shortening and death of the lineage would result in the extinction of the species. Furthermore, natural selection operates directly on the yeast cell, while it is filtered by the organism in the case of an individual metazoan cell which can perform in an environment that has its own selective exigencies. Old yeasts and old animals have "escaped the force of natural selection". Any postreproductive advantages or disadvantages are not selected upon directly to appear in the progeny. This is obvious for old animals; we shall see that it is also true for yeast.

How do we define yeast aging? Vegetative yeast cells, not to mention meiotic spores, can remain in a nondividing state for long periods of time. In this condition, they metabolize very slowly and do not produce progeny. Although of considerable importance, this does not resemble the situation of a typical mammal during its life span. We therefore turn to a population of growing and dividing yeast cells to search for a model of mammalian aging. Unfortunately, we cannot easily define the physiological age of a particular yeast by any set of known parameters. It is far better to draw conclusions regarding variables that influence yeast aging by examining their effect on life span. Thus, longevity becomes a tool in the study of aging. Fortunately, yeasts have rather ephemeral life spans of only a few days. From what has been said, it should be apparent that the measure of life span is the number of divisions or generations an individual cell completes, which is the equivalent to the number of progeny produced by an individual yeast cell. Indeed, Mortimer and Johnston (1959) observed that individual cells divided only a limited number of times. Muller et al. (1980), in turn, demonstrated that this replicative life span, rather than chronological life span, is characteristic for a yeast strain. This is because temperature and the availability of nutrients can affect the rate of cell growth and division, altering chronological life span, but leaving replicative life span immutable. Having said this, it is necessary to point out that longevity is a statistical phenotype in yeast as in other species. The mean life span is a characteristic feature of a particular yeast strain (Egilmez and Jazwinski 1989), indicating that it is genetically determined.

From this discussion, it follows that, for the study of aging, yeasts have an advantage in addition to their genetics. This advantage is their short life span of only a few days. The methodology for studying yeast aging has been developed (Kim et al. 1998). This includes procedures for the preparation of old yeast cells for molecular and biochemical analysis.

2
Phenomenology of Yeast Aging

The phenomenology of yeast aging has been reviewed (Jazwinski 1990a,b, 1993, 1996a, 1999a; Sinclair et al. 1998a). Table 1 summarizes the plethora of morphological and physiological changes that yeasts suffer as they proceed through their replicative life spans. Is this pathophysiology of yeast aging informative of its cause or causes? This question is difficult to answer at present. Cause and effect, as well as compensatory or coincidental changes, are likely to reside in the list. It should be noted that the age changes are quite individual, and the listing represents an average over the entire population. An excellent example of this are changes in cell polarity, as manifested by the occurrence of random budding (Fig. 1). Overall, there is a substantial loss of polarity in an aging cohort (Jazwinski et al. 1998). If there is one change that may be universal in its occurrence, pattern, and magnitude, it is the increase in generation time with age.

Apart from age changes, the other hallmark of organismal aging is an exponential increase in mortality rate with age. Such an increase is found in yeast (Pohley 1987; Jazwinski et al. 1989). This increase in mortality rate signifies that the probability that a yeast cell continues dividing decreases exponentially as a function of the number of divisions already completed. Recently, it has been shown that mortality rate plateaus at later ages in yeast (Jazwinski et al. 1998), as it does in fruit flies (Curtsinger et al. 1992) and worms (Vaupel et al. 1994). This plateau is readily explained by the epigenetic stratification of the yeast population due to individual change (Jazwinski et al. 1998; Jazwinski 1999b).

The proportion of old cells of a given age in a yeast population declines exponentially as a function of their age, because with each division a young progeny cell is produced. A 20-generation old cell will constitute only one part in about one million. Additionally, this old cell will divide at a rate considerably slower than the division rate of a young cell (Egilmez and Jazwinski 1989). As a result, the old cell contributes little to the evolution of

Table 1. Yeast age changes

Characteristic[a]	Change
Cell size	Increase
Cell shape	Altered
Granular appearance	Develops
Surface wrinkles	Develop
Loss of turgor	Develops
Cell fragility (prior to death)	None
Cell lysis	Occurs
Bud scar number	Increase
Cell wall chitin	Increase
Vacuole size	Increase
Generation (cell cycle) time	Increase
Response to pheromones (haploids)	None[b]
	Decrease[c]
Mating ability (haploids)	Decrease
Sporulation ability (diploids)	Increase
Cell cycle arrest at G_1/S boundary (putative)	Occurs
Senescence factor	Appears
Mutability of mtDNA	Decrease
UV resistance	Increase/decrease
Telomere length	None
Random budding[d]	Increase
Specific gene expression	Altered
rRNA levels	Increase
rDNA circles[e]	Increase
Cellular rRNA concentration[f]	Decrease
Protein synthesis	Decrease
Ribosome activity, polysome recruitment	Decrease
Transcriptional silencing[g]	Decrease
Size difference between mother and daughter[h]	Decrease

[a]References provided in Jazwinski (1996a), except as noted; [b]Muller (1985); [c]Smeal et al. (1996); [d]Jazwinski et al. (1998); [e]Sinclair and Guarente (1997); [f]Due to increase in cell size; [g]Kim et al. (1996), Smeal et al. (1996); [h]Egilmez and Jazwinski (1989), Kennedy et al. (1994).

the species. Natural selection has little opportunity to operate on the old cell to affect the fitness of the population.

Each time a yeast divides it is marked by a ring of chitin-containing material, called the bud scar (Bartholomew and Mittwer 1953; Cabib et al. 1974). These bud scars can be readily visualized, and they constitute a marker for the nominal age of the yeast cell. The nominal age, as opposed to the physiological age, does not tell us what the probability is that an individual cell will die. It is difficult to count the number of bud scars on a cell when they constitute more than a handful. The occupation of the cell surface by a cumulative number of bud scars was recognized early on as a potential cause of yeast aging (Mortimer and Johnston 1959). There are strong arguments, both logical and experimental, that they are not (reviewed recently by Jazwinski 1996a). Most tellingly, it is possible to extend the life span of yeasts by both genetic and environmental manipulation, as will be seen below, indi-

Fig. 1. Age changes show variability between individual yeasts. Cell polarity changes were determined for individual yeasts during their life spans. The budding patterns were scored as axial, random, or unknown for randomly selected *S. cerevisiae* SP1-1 cells as a function of age (Jazwinski et al. 1998). Three of these are shown here. The patterns of cell polarity change were unique for each yeast in the 200 cells examined. *Unknown* indicates the inability to score a particular budding because the previous bud had detached from the mother. (Data courtesy Dr. Chi-Yung Lai in my laboratory)

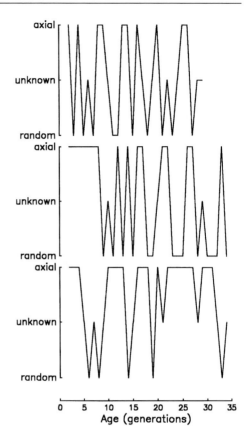

cating that the bud scar, in and of itself, is not limiting for longevity. It is, however, possible that there is some upper limit to the number of bud scars a yeast cell can accumulate; but, this limit has not been reached thus far, in any experiment.

3
Genetics of Longevity

The phenomenological description of yeast aging set the stage for a genetic dissection of the aging process. The past decade has seen the identification of a dozen genes that play a role in determining yeast life span (Table 2). Some of these genes were new, while for others a role in aging was a new phenotypic assignment. Three routes have resulted in the discovery of these longevity genes. The first was the consequence of the cloning of genes that are differentially expressed during the yeast life span (Egilmez et al. 1989). The

Table 2. Yeast longevity genes

Gene	Function	Reference
CDC7	Protein kinase; cell cycle control and transcriptional silencing	Egilmez and Jazwinski (1989); Jazwinski et al. (1989)
LAG1	Endoplasmic reticulum (ER) protein; GPI-anchored protein transport	D'mello et al. (1994)
LAC1	Homologue of LAG1; ER protein; GPI-anchored protein transport	Jazwinski (1996b); Jiang et al. 1998
PHB1	Unknown; mitochondrial protein	Jazwinski (1996b); Coates et al. (1997); Berger and Yaffe (1998)
PHB2	Unknown; mitochondrial protein; homologue of PHB1	Coates et al. (1997); Berger and Yaffe (1998)
RAS1	GTP-binding (G) protein; involved in signal transduction	Sun et al. (1994)
RAS2	G-protein; involved in signal transduction	Sun et al. (1994)
RTG2	Unknown; effector of retrograde response	Kirchman et al. (1999)
SIR4	Transcriptional silencing	Kennedy et al. (1995)
SGS1	DNA helicase; DNA recombination	Sinclair et al. (1997)
UTH4	Unknown	Kennedy et al. (1995, 1997)
YGL023	Unknown; homologue of UTH4	Kennedy et al. (1997)

second was the selection of mutants based on secondary phenotypes that were expected to be associated with enhanced survival (Kennedy et al. 1995). The third derived from the analysis of processes that determine yeast life span (Sun et al. 1994; Sinclair et al. 1997; Kirchman et al. 1999).

The first yeast longevity gene to be described was *LAG1* (D'mello et al. 1994). Yeast possess a homologue of this gene, called *LAC1* (Jazwinski 1996b; Jiang et al. 1998). The human and *C. elegans* homologues of this gene have been cloned (Jiang et al. 1998). The human homologue can perform longevity gene function in yeast. Recently, it has been shown that the Lag1 and Lac1 proteins facilitate the transport of GPI (glycosylphosphatidylinositol)-anchored proteins (Barz and Walter 1999). *LAG1* is a gene that is differentially expressed during the yeast life span. The involvement of the yeast *RAS* genes in determining longevity was already indicated by the extension of life span by an activated allele of mammalian *ras* (Chen et al. 1990). The yeast *RAS* genes impact at least three signal transduction pathways (Tatchell 1993; Kaibuchi et al. 1986; Mosch et al. 1996) to modulate a variety of physiological responses (Marchler et al. 1993; Tatchell 1993). As discussed later, their effects on yeast longevity appear to be quite pleiotropic. The *SIR4* gene was implicated in yeast aging through the life extension afforded by a gain-of-function mutant, *SIR4-42* (Kennedy et al. 1995). The Sir4 protein is a component of the Sir silencing complex, which represses transcription in heterochromatic regions of the yeast genome (Loo and Rine 1995). Silencing of two of the three heterochromatic regions, the silent mating-type loci and subtelomeric genes, is known to decline with aging (Kim et al. 1996; Smeal et al. 1996).

Even a cursory examination of the list of genes provided here makes the notion of a single pathway determining life span seem rather unlikely. These genes appear to be involved in a rather diverse array of processes. The existence of more than one pathway in life-span determination conjures up the idea of more than one mechanism of aging in yeast. Evidence for at least two mechanisms has already been adduced, and this is likely not to be the end of the story, as we shall see below.

The term longevity determination or determining might suggest a deterministic or programmed process. There is, however, no evidence at this time to invoke such a program operating from birth to death. Furthermore, the existence of such a program is not consistent with evolutionary theory (Finch 1990; Rose 1991). A genetic program is one in which there is a predictable series of genetically determined events, each one dependent on the completion of the former in the series. Aging is a stochastic process (Finch 1990; Rose 1991; Jazwinski et al. 1998). The genetic constitution of the individual organism determines the bounds within which aging plays itself out in interactions with the environment and due to other epigenetic factors. Indeed, the genotype and development establish the life maintenance reserve, which is an abstract concept that denotes the sum of the interactions of genetic, environmental, and epigenetic factors (Jazwinski 1996b). The life maintenance reserve determines the functional potential that facilitates the survival of the organism to reproductive maturity.

4
Physiological and Molecular Mechanisms of Aging

4.1
Genetic Instability and Gene Dysregulation

The role of genetic stability in yeast aging has been discussed in detail (Sinclair et al. 1998b), thus only a cursory account is provided here. Old yeast cells have a high content of circular DNA species (ERC), apparently resulting from the excision of ribosomal DNA repeats and their amplification (Sinclair and Guarente 1997). This is the manifestation of genetic instability that has been recognized in this organism. The ERCs are present as monomers as well as multimers. These ERCs can be generated artificially, and this is associated with a reduction of life span. More significantly, the partial inactivation of the Cdc6 protein, which is involved in initiation of DNA replication and would result in a decrease in the propagation of the circles, extends life span. The

ERCs segregate preferentially to the mother cell during cell division. Thus, accumulation of ERCs is associated with normal aging, and this accumulation appears to limit longevity. The possible pleiotropic effects of the inactivation of Cdc6 protein are a frequently encountered problem in aging studies. They warrant some caution in the interpretation of the results.

The deletion of the *SGS1* gene, which is involved in suppressing recombination at the ribosomal DNA locus, results in an increase in ERC production and a concomitant decrease in life span (Sinclair and Guarente 1997; Sinclair et al. 1997). This curtailment of longevity has been interpreted as premature senescence in yeast. This characterization is of great interest because *SGS1* is a homologue of the human *WRN* gene, which is mutated in the segmental progeroid syndrome called Werner's syndrome (Yu et al. 1996). The conclusion that the *sgs1* mutant yeasts undergo premature senescence is based on the observation that they precociously lose the ability to respond to the mating pheromone α-factor, indicating that they are functionally diploid while remaining genetically haploid. The loss of response to the pheromone is a feature of normal aging in yeast, and it is the result of the loss of silencing at the silent mating type loci (Smeal et al. 1996) due to the migration of Sir complexes to the nucleolus (Kennedy et al. 1997; Sinclair et al. 1997). The production and amplification of ERCs, which the *sgs1* mutation triggers, provides the substrate to which the Sir complexes migrate, and it can be considered the direct cause of the loss of silencing. Because the *sgs1* mutation is the direct source of the specific aging phenotype (loss of response to pheromone) that is used as an indicator of premature senescence, the loss of response to α-factor cannot be ascribed to the acceleration by this mutation of the global aging process of which it is only a part. This quandary represents one of the forms in which the difficulty of interpretation of life-span-shortening treatments, as opposed to extension of life span, manifests itself.

The studies on ERCs demonstrate the significance of genetic instability in yeast aging. The nucleolar enlargement and fragmentation found in old yeast cells (Kennedy et al. 1997) can be readily explained by the appearance of ERCs. It has been suggested that the migration of silencing proteins from the telomeres and the silent mating-type loci is the result of aging but that it can also retard aging (Kennedy et al. 1997). The molecular basis for these effects is not clear at present. The chromosomal rDNA locus has been proposed to constitute the AGE locus postulated in earlier studies (Kennedy et al. 1995). This proposal is based on the fact that *SIR4-42* extends life span (Kennedy et al. 1995) and, at the same time, causes migration of the Sir complex to the nucleolus (Kennedy et al. 1997). It would be interesting to see whether *SIR4-42* mutants exhibit a delay in ERC production and nucleolar fragmentation. The question of how Sir complex migration actually elicits its effect on aging is open.

Another form of instability may also be involved in yeast aging. This instability is epigenetic in nature and has been termed gene dysregulation (Jazwinski 1996b). It has been shown that there is a loss of chromatin-dependent transcriptional silencing in two known heterochromatic regions of the yeast genome, the subtelomeric loci (Kim et al. 1996) and the silent mating-type loci (Smeal et al. 1996). The *SIR4* and the *CDC7* genes affect silencing (Laurenson and Rine 1992), and they are involved in determining life span (Egilmez and Jazwinski 1989; Kennedy et al. 1995). It has also been shown that *RAS2*, another longevity gene in yeast, affects silencing of subtelomeric genes (Jazwinski et al. 1998). These studies all suggest a role for stability of epigenetic mechanisms of gene regulation in determining life span. However, the interpretation of the results is complicated, again due to the pleiotropic nature of the genes involved. The enhanced stress resistance of the *SIR4-42* mutant (Kennedy et al. 1995) is likely to have an important salutary role on longevity. However, this mutant also induces migration of Sir complexes to the rDNA locus (Kennedy et al. 1997) and any effects this might have on yeast longevity. The life-extending effect of *SIR4-42* is a special case, because mutants in other *SIR* genes do not show extended life span.

4.2
Metabolic Control

The measure of the yeast life span is the number of daughters an individual cell produces. The production of these daughters requires substantial metabolic activity in the form of biosyntheses and energy generation. Thus, the measure of the yeast life span is actually total metabolic capacity. This is important to note, because life extension in yeast, by definition, involves enhancement of total metabolic capacity. This feature distinguishes yeast aging studies from similar studies in other organisms, in which the question may arise whether life extension may be due to reduction in metabolic rate because the measure of life span is chronological.

There is now more direct, mechanistic evidence that metabolic activity and its control play a major role in determining yeast life span (Kirchman et al. 1999). This stems from the observation that the life span of a petite yeast, which does not possess fully functional mitochondria, is substantially longer than that of its parent grande strain. This extension of life span does not appear to be due to a reduction of oxidative stress under the conditions in which the yeasts are grown. Rather, it appears to be due to the induction of the retrograde response, which is an intracellular signaling pathway from the mitochondrion to the nucleus that results in the induction of a variety of nuclear genes (Parikh et al. 1987). The life extension or life-span maintenance is correlated with the retrograde response in four different yeast strains. It is suppressed by the deletion of a gene, *RTG2*, which encodes a downstream effector of the retrograde response (Kirchman et al. 1999). The Rtg2 protein has been proposed to facilitate the formation of an active heterodimer of two other downstream effectors of the retrograde response (Rtg1 and Rtg3 proteins), which form a transcription factor that induces gene expression from the retrograde response element found in the promoters of certain nuclear genes (Rothermel et al. 1997). The retrograde response can be manipulated both genetically and environmentally to extend yeast life span (Kirchman et al. 1999), providing strong evidence for a molecular mechanism of aging.

The Ras2 protein plays a role in nutritional assessment in yeast (Tatchell 1993). It is therefore not surprising that the *RAS2* gene impinges upon the retrograde response. Deletion of this gene abrogates the life extension seen in the petite, and it modulates the expression of the nuclear genes that are induced in the retrograde response (Kirchman et al. 1999). Our current understanding of the role of the retrograde response in determining yeast life span is summarized in Fig. 2. This figure also lists the nuclear genes that are regulated by the Rtg1-3 proteins. These include metabolic enzymes that are found in mitochondria, peroxisomes, and in the cytoplasm (Chelstowska and Butow 1995; Small et al. 1995). The effect of induction of the retrograde response, judging by the enzymes induced, represents a shift from the utilization of glucose to acetate, activation of gluconeogenesis, and stimulation of the glyoxylate cycle to generate Krebs cycle intermediates.

The physiological effect of induction of the retrograde response is an adaptation to survival on reduced energy and carbon sources. This response appears to signal changes in mitochondrial activity. Some of the metabolic changes observed in the retrograde response are also observed during yeast sporulation, which constitutes a response to severe caloric stress. The retrograde response documents the importance of metabolic control in determining longevity. It has been found that induction of the response results in a dramatic increase in ERCs (Conrad-Webb and Butow 1995). This indicates that the retrograde response predominates in the extension of life span even in the presence of ERCs.

Yeasts are a particularly useful model for the analysis of metabolic factors in aging. They can do without mitochondrial respiration, which allows one to separate metabolic control from the stress associated with the production of oxidants. Related to this is the fact that yeasts can use glycolysis as their sole energy source, so the metabolic factors associated with respiration can be considered separately. While the regulation of mitochondrial metabolism in *S. cerevisiae*, a facultative aerobe, and in other organisms typically used to study aging, which are all obligate aerobes, must have some fundamental differences, these differences may not be as striking as the similarities. The primary apparent distinction between yeast and these organisms is the capacity of yeast to ferment. This differs from anaerobic glycolysis found in tissues such as skeletal muscle only by the end product, which is ethanol in yeast and lactate in muscle. Furthermore, petite animal cells that lack fully functional mitochondria can be maintained on glucose in tissue culture (King and Attardi 1989), blurring the distinction between yeast and animal cells. The ability to generate and manipulate petite yeast allows one to perturb

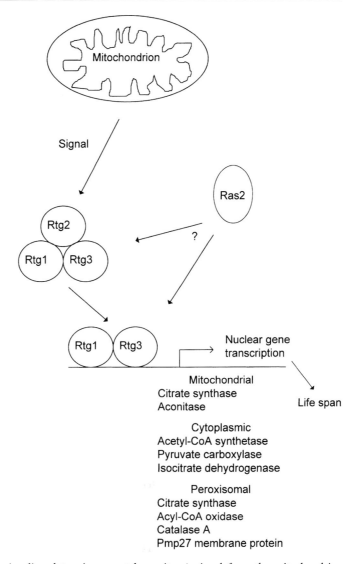

Fig. 2. Intracellular signaling determines yeast longevity. A signal from the mitochondrion leads to the activation of transcription of nuclear genes by the Rtg1-Rtg3 protein heterodimer. The Rtg2 protein has been postulated to facilitate the formation of the Rtg1-Rtg3 protein complex. The identity of the signal generated by the mitochondrion is not known. The nuclear genes controlled by the *RTG* genes encode proteins found in the mitochondria, cytoplasm, and peroxisomes. It is likely that there are many more than those listed here. Activation of this retrograde response pathway leads to life extension. It is not known which of the targets of this pathway determine yeast life span. *RAS2* interacts with this pathway somewhere downstream of the mitochondrial signal, and it is likely that it determines life span by mechanisms in addition to the retrograde response

mitochondrial metabolism to more easily uncover certain metabolic relationships, which would be more difficult to discern in obligate aerobes. Although precise molecular details may be different, the broad physiological
principles gleaned from such studies may be the same in yeast and mammals.

4.3
Stress Resistance

Dwindling nutritional resources and frank starvation result in stress in yeast,
with sporulation being the extreme response. This form of stress has been
touched upon in the discussion of metabolic control. Yeasts are also subjected to a variety of other environmental stresses. One of these is ultraviolet
radiation (UV). Resistance to UV actually increases in yeast until mid-life,
after which it declines rapidly (Kale and Jazwinski 1996). This biphasic response is closely paralleled by the expression of the *RAS2* gene. This gene is
essential for resistance to UV (Engelberg et al. 1994). Part of its role in
determining yeast life span may involve resistance to this stress, when it is
encountered. There are other reasons to connect stress resistance to longevity
in yeast. Selection for resistance to cold stress and starvation resulted in the
isolation of mutants displaying extended life span (Kennedy et al. 1995).
These long-lived mutants were resistant to other stresses as well.

Yeasts also encounter heat stress in the wild. Normally, such episodes are
sublethal. Chronic exposure of yeasts to repeated, sublethal heat stress
shortens their life span (Shama et al. 1998a). This life-shortening effect is
exacerbated in a *ras2* mutant, but not in a *ras1* mutant, indicating that *RAS2*
exerts a protective effect under these conditions (Shama et al. 1998a). Cells
mutated for *RAS2* display a pronounced delay in the resumption of growth
and division after a bout of sublethal heat stress. This stress results in a rapid
induction of stress genes, such as *HSP104*, and repression of growth-promoting genes, such as *CLN2*. However, there is a marked delay in the downregulation of the stress genes and in the upregulation of the growth-promoting
genes in the *ras2* mutant, whereas there is no difference between *ras1* and wild
type (Shama et al. 1998a). Thus, the process of recovery from the stress is
defective in *ras2*. Many stress genes in yeast possess the stress response-
regulatory element (STRE) in their promoters. *RAS2* modulates expression
from the STRE in a negative manner (Marchler et al. 1993). Other forms of
stress are related to the heat-shock response. The major heat-shock protein in
yeast, Hsp104, is induced by a variety of stresses (Sanchez et al. 1992), and
oxidative stress plays a role in heat-induced cell death (Davidson et al. 1996).

The life-span curtailment on episodic exposure to sublethal heat stress is
rescued by overexpression of *RAS2* (Shama et al. 1998a). This overexpression
not only overcomes the deficit, but it actually extends the life span beyond
that of unstressed control cells. This effect is very similar to that observed

when *RAS2* is overexpressed in the absence of any overt stress (Sun et al. 1994). There is a difference, though. In the absence of overt stress, the life extension occurs through a cAMP-independent pathway (Sun et al. 1994), while the Ras2-cAMP pathway is essential for the recovery and life extension observed during chronic stress (Shama et al. 1998a).

Unlike chronic stress, transient exposure of yeast to sublethal heat stress results in increased longevity (Shama et al. 1998b). This effect is due to a pronounced reduction in mortality rate that persists for several generations but is not permanent. This suggests a heritable epigenetic change in the cells, the basis of which is not known at present but could include alterations in chromatin structure. For this life extension, both of the *RAS* genes are required, distinguishing the phenomenon from the effects of chronic stress. *RAS2* is necessary for the timely recovery from the transient stress just as it is during chronic stress events, but *RAS1* must play a different role because it cannot substitute for *RAS2* in this regard due to its poor stimulation of adenylate cyclase. The *HSP104* gene is essential for life extension by transient heat stress (Shama et al. 1998b). This gene encodes the major heat-shock protein that is largely responsible for induced thermal tolerance in yeast (Lindquist and Kim 1996). This life extension also requires functional mitochondria (Shama et al. 1998b). Petite yeasts exposed to transient, sublethal heat shock do not exhibit life extension but rather the synthetic phenotype of life-span shortening, suggesting an interaction between the life-extension mechanisms activated by transient heat stress and by the retrograde response (Shama et al. 1998b). Both mechanisms might not operate concurrently. In fact, petites are more resistant to lethal heat shock (C.-Y. Lai and S. M. Jazwinski, unpubl.). This could be due to induction of heat-shock proteins in petites, as has been found in animal cells that have lost their mitochondrial DNA (Martinus et al. 1996).

The studies on the role of the *RAS* genes in the response of yeast to heat stress have revealed some of the functional differences between them, but they have also shown the intricate nature of the role of these genes in maintaining homeostasis. Our understanding of the role of *RAS1* and *RAS2* in the response to heat stress is shown in Fig. 3. This model has been confirmed experimentally in many of its features. The proposed role of *RAS1* in the long-term effects of transient heat stress remains to be documented. This gene cannot substitute for *RAS2* in the response to chronic heat stress, most likely due to its weak stimulation of adenylate cyclase. Instead, the model proposes that it is the stimulation of inositolphospholipid turnover promoted by *RAS1* (Kaibuchi et al. 1986; Hawkins et al. 1993) which is responsible. *RAS1* might cause a change in inositol lipids that survives for many generations, just like the decrease in mortality rate. In yeast, sphingolipids contain inositol, and they are required for resistance to a variety of stresses, among them heat (Patton et al. 1992). The question might be asked why yeasts do not simply upregulate stress responses sufficiently to allow them to survive a

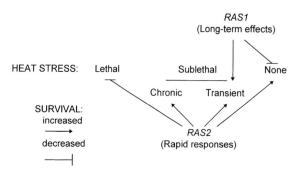

Fig. 3. Heat stress and the role of the *RAS* genes in determining yeast life span. *RAS2* modulates the expression of stress genes. It prevents expression of these genes at levels sufficient to withstand an acute (lethal) heat shock and reduces survival in stationary phase (Tatchell 1993). *RAS2* aids rapid recovery from sublethal heat stress, whether transient or chronic, and enhances life span by facilitating downregulation of the expression of stress genes. It also promotes longevity in the absence of overt stress. Activation of a cAMP-independent pathway is responsible for this effect, while in the presence of heat stress the cAMP pathway is also required. The cAMP-independent pathway is likely to be a mitogen-activated protein kinase (MAPK) pathway (Mosch et al. 1996). *RAS1* has a life-span-shortening effect in the absence of stress, while having no effect on survival in the presence of chronic, sublethal stress. This gene is required for the life extension effected by transient, sublethal heat stress, postulated to be the effect of its stimulation of inositol phospholipid (PI) turnover

lethal heat shock. It would appear that they trade off this capacity in order to be able to live a normal life span in the face of the more commonly encountered chronic heat stress.

4.4
Coordination of Metabolic Activity and Resistance to Stress

It is clear that growth and division, which are necessary for yeasts to progress through their life span, are temporarily halted during heat stress. This implies coordination. Environmental stress, as well as stress generated within the cell, creates an ever-changing backdrop to the activities required for survival and a normal life span. In some cases, these activities would be incompatible with the genetic and epigenetic exigencies to which the cell is exposed. Depending on the environmental and epigenetic factors at play, the optimal allocation of cellular resources would differ. The expression of an activated allele of *RAS* extends life span (Chen et al. 1990). However, an increase in this expression beyond a certain level abrogates this effect. The response is biphasic. This suggests that there is an optimal level of expression that results in a maximal increase in life span, under constant conditions. The effect of transient, sublethal heat stress on yeast longevity is salutary (Shama et al. 1998b), but when it becomes chronic it is deleterious (Shama et al. 1998a). In fact, this effect of sublethal heat stress is biphasic (S. Shama, C.-Y. Lai, and

S. M. Jazwinski, unpubl.). With increasing frequency of heat-stress episodes, there is first a progressive increase in the extension of life span, followed by a decline and a net reduction in longevity. Apparently, this is a nonlinear process. In this case, the expression of *RAS* is the constant, and the environmental input changes.

The *RAS* genes constitute a homeostatic device in yeast longevity (Fig. 4). They appear to perform in this capacity by modulating a wide array of processes that are crucial for yeast longevity. The role of these genes in metabolic control and in stress responses has been discussed above. These genes also interact with pathways signaling cell division (Baroni et al. 1994; Sun et al. 1994; Morishita et al. 1995). *RAS2* modulates cellular spatial organization by moderating the budding pattern (Jazwinski et al. 1998), and it potentiates pseudohyphal growth in yeast, which involves a distinct change in the budding pattern (Mosch et al. 1996). It is possible that genetic stability is also modulated by the *RAS* genes. Cellular resources, including nutrients and energy, must be allocated to the competing demands of all of the processes listed in Fig. 4 and others that may not be recognized at present. All of these processes are important for life maintenance and longevity. The *RAS* genes, as the homeostatic device in yeast longevity, modulate these processes and thereby indirectly accomplish the apportionment of resources. The only signal to which the *RAS* genes appear to respond is nutritional status (Tatchell 1993).

The intracellular environment of the yeast cannot be a constant. It must fluctuate in order to accommodate temporal changes in the allocation of the cellular resources due to changing demands. The *RAS* homeostat cannot therefore be locked into one setting, but it must be able to establish an equilibrium whose set point can change with time. Random events and the process of change, including set point change, might throw the homeostat off. Even if this shift is small, there may exist a finite probability that the homeostat would find itself beyond its "normal" range. This would establish a new state with a new range of set points for the homeostat. This new state

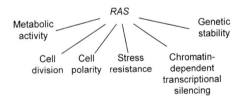

Fig. 4. The *RAS* genes maintain homeostasis during the yeast life span. This homeostasis is maintained through the coordination of cellular activities. These genes can affect so many cellular functions, because they are not integral components of the pathways that regulate them. Instead, the *RAS* genes modulate these pathways. In this way, they figuratively channel resources to the competing demands of these cellular functions. This occurs under conditions of shifting internal and external environment. The only known signal to which *RAS1* and *RAS2* respond is nutritional status

might be even more prone to change than the former one. The external environment of the cell places additional pressure on the homeostat, by accelerating and amplifying the changes it needs to accommodate.

There are many changes that occur during yeast aging (Table 1). These changes may, in the first instance, be more of a cause than a result of aging. The changes in cell polarity during aging are apparently random, and they seem benign (Fig. 1). Yet, they place pressure on the *RAS* homeostat to accommodate them, thereby altering its set point. This process would lead to the loss of homeostasis and result in aging. A formal model of the role of individual change in the biological aging process has been presented, and some of its predictions have been tested (Jazwinski et al. 1998; Jazwinski 1999b).

4.5
Comparisons with Other Organisms

The genetic dissection of aging in the nematode *C. elegans* and in the fruit fly *Drosophila* has provided interesting correlates with two of the yeast molecular mechanisms of aging – metabolic regulation and stress resistance. The nematode *daf* genes describe a developmental pathway for the formation of a dispersal form called the dauer larva, not unlike a yeast spore. The environmental triggers that result in dauer formation are starvation, crowding, and heat. Certain mutants in one branch of this pathway influence adult longevity, and the genes encode components of an insulin-like signaling pathway (Tissenbaum and Ruvkun 1998). Genetic manipulation of this pathway can double the life span.

The *daf-2* gene codes for an insulin/IGF-like receptor (Kimura et al. 1997) and lies at the head of this pathway (Kenyon et al. 1993; Larsen et al. 1995). The downstream effector of this pathway is the *daf-16* gene, whose product is a transcription factor belonging to the forkhead family (Ogg et al. 1997; Lin et al. 1997). In mammals, insulin regulates metabolism through the transcription factor HNF-3, which regulates the expression of metabolic genes such as the gluconeogenic enzyme phosphoenolpyruvate carboxykinase and the GLUT4 glucose transporter (Lai et al. 1990; O'Brien et al. 1995). In the nematode, *daf-16* affects glycogen and fat storage (Ogg et al. 1997). The other known components of this pathway that lie between *daf-2* and *daf-16* are *age1/daf-23*, which encodes a phosphatidylinositol-3-OH kinase (Morris et al. 1996), *daf-18*, whose product is a homologue of the human tumor suppressor PTEN that possesses phosphatidylinositol 3,4,5-triphosphate phosphatase activity (Ogg and Ruvkun 1998), and *akt-1* and *akt-2*, which code for nematode homologues of the Akt/PKB protein kinase (Paradis and Ruvkun 1998). One other gene, *daf-12*, which encodes a zinc finger protein belonging to the thyroid hormone/steroid/retinoic acid receptor superfamily, interacts with this pathway (Larsen et al. 1995). The activation of the *daf-2/daf-16* pathway leads to changes in metabolic enzyme activity that are reminiscent of those seen in the retrograde response, including a shift to the glyoxylate cycle (Vanfleteren and De Vreese 1995).

There is a second pathway that determines the longevity of *C. elegans*. It is described by the *clk-1* gene, which is one of a group of three clock genes (Wong et al. 1995; Lakowski and Hekimi 1996). These clock genes affect a variety of processes, including growth rate. The *clk-1* gene encodes a homologue of the yeast *CAT5/COQ7* gene (Ewbank et al. 1997), which is a regulator of mitochondrial function. The *age-1/daf-23* mutant is more resistant to oxidative stress, and this is associated with increased levels of catalase and Cu,Zn-superoxide dismutase (Larsen 1993; Vanfleteren 1993). The long-lived *daf* mutants and the *clk-1* mutant are more resistant than the wild type to UV, heat, and oxidative stress (Murakami and Johnson 1996). The studies with the *daf* and *clk-1* mutants point to the importance of metabolic regulation and stress resistance for longevity, and they suggest an association between these physiological functions. Further support for the role of stress resistance comes from studies in which the induction of thermal tolerance extended the nematode life span (Lithgow et al. 1995).

Fruit flies displaying an extended life span have been obtained by selection for delayed reproduction (Luckinbill et al. 1984; Rose 1984). These flies have higher glycogen and lipid stores (Service et al. 1985; Service 1987; Graves et al. 1992; Arking et al. 1993; Dudas and Arking 1995). Associated with these metabolic changes are alterations in glucose-6-phosphate dehydrogenase isoforms (Luckinbill et al. 1989), which would certainly favor fat synthesis and storage. The flies utilize nutrients more efficiently than the control flies (Riha and Luckinbill 1996), and they have an expanded metabolic capacity, as measured by lifetime oxygen consumption (Arking et al. 1988). In the case of yeast, nematodes, and fruit flies, the metabolic adaptations associated with extended longevity appear to represent a shift to "life on acetate", with the utilization of fatty acids as a source of energy and the activation of gluconeogenesis. In yeast and nematodes, there is also an activation of the glyoxylate cycle to provide Krebs cycle intermediates.

The fruit flies selected for an extended life span are more resistant to starvation, dessication, heat, and ethanol than the controls (Service et al. 1985; Service 1987; Graves et al. 1992). Additional gains in longevity were observed in these flies when they were selected further for starvation and dessication resistance (Rose et al. 1992). Certain lines of these flies showed enhanced resistance to oxidative stress, and this was associated with increased levels of antioxidant enzymes (Arking et al. 1993; Dudas and Arking 1995). The life span of transgenic flies expressing Cu,Zn-superoxide dismutase (Parkes et al. 1998; Sun and Tower 1999), or the dismutase together with catalase (Orr and Sohal 1994), was extended. The induction of thermal tolerance in fruit flies also extends their life span (Khazaeli et al. 1997), and this increase is proportional to the dosage of the Hsp70 gene (Tatar et al.

1997). In addition, a *Drosophila* mutant that was isolated for extended life span is resistant to a variety of stressors (Lin et al. 1998).

Caloric restriction of mammals results in extended longevity (reviewed in Masoro 1995). It involves a reduction in the caloric content rather than a change in the composition of the diet. In comparison, the retrograde response appears to constitute an adaptation to a shift from a nutrient of high caloric content (glucose) to one of lower caloric content (acetate). Calorically restricted animals show a variety of metabolic changes, such as reduced blood glucose and insulin levels (reviewed in Masoro 1995). They have higher insulin receptor levels (Pahlavani et al. 1994) and enhanced activity of the gluconeogenic enzyme phosphoenolpyruvate carboxykinase (Van Remmen et al. 1994). Caloric restriction, furthermore, results in the maintenance of antioxidant enzyme levels late in life (Xia et al. 1995). It also results in the enhancement of resistance to heat stress in older animals (Heydari et al. 1993). Caloric restriction leads to an elevation in circulating glucocorticoids (Sabatino et al. 1991). This might be interpreted as a response to sublethal stress that has a life span-extending effect similar to the effects of heat stress in yeast, nematodes, and fruit flies.

5
Primacy of Metabolic Control

The discussion of the role of metabolic regulation and of stress resistance in determining life span raises the question of which of these physiological processes is primary. The fact that caloric restriction has significant effects on antioxidant enzyme levels, resistance to heat, and stress hormones suggests that metabolic control may be the primary effector, while stress resistance follows in tow and compensates. Certainly, metabolism, and in particular oxidative metabolism, generates much cellular stress, which must be mitigated for the metabolic activity necessary for life maintenance to operate. In many tissues, this mitigation may be achieved through the induction of antioxidant enzyme activities (Xia et al. 1995). In others, such as skeletal muscle, it may be the reduction of the generation of oxidative stress, as deduced from an attenuation of the age-related increase in antioxidant enzyme activities on dietary restriction (Luhtala et al. 1994). The proximal cause of this attenuation may be the reduction in oxidative stress due to more efficient functioning of the mitochondria.

There are other arguments in favor of the primary role of metabolic control. The Rtg2 protein downstream effector of the retrograde response in yeast upregulates expression of the aconitase gene (Velot et al. 1996). Because this mitochondrial enzyme is a specific target of oxidative damage during aging (Yan et al. 1997), this activation could be interpreted as an enhancement of resistance to stress. The retrograde response also upregulates the expression of the peroxisomal catalase gene, *CTA1* (Chelstowska and Butow 1995). The *RAS2* gene is a positive modulator of the retrograde response (Kirchman et al. 1999). The recovery from glucose starvation requires *RAS2* (Tatchell 1993), in similarity to the protective effect of this gene under conditions of chronic, sublethal heat stress (Shama et al. 1998a). *RAS2* is required for recovery from the stress (Shama et al. 1998a), and this function is effected through downregulation of expression from the STRE (Marchler et al. 1993). These observations, coupled to the increase in viability on lethal heat stress in a *ras2Δ* (Tatchell 1993), suggest that yeasts have two fundamental responses to heat stress. They can either "stay and fight" by copious induction of their stress responses, or they can "run away", so to speak, by rapid growth and division to increase the number of progeny, some of which can survive. As pointed out earlier, the responses to various modes of stress are related in yeast. The most obvious distinction between them is whether they are environmental or intrinsically generated.

The *SNF1* protein kinase gene is required to release yeast from glucose repression when the cells are starved for glucose (reviewed in Johnston 1999). *SNF1* is required for survival in stationary phase. It is also important in activating protective systems against other forms of stress through the expression of genes that are glucose-repressible (Alepuz et al. 1997). The *SNF1* homologue is also involved in stress responses in mammalian cells (Hardie and Carling 1997). All of these facts suggest a strict relationship between metabolic control and stress responses, and they provide some support for the notion that metabolic control has primacy.

Acknowledgments. The research in the author's laboratory is supported by grants from the National Institute on Aging of the National Institutes of Health (USPHS).

References

Alepuz PM, Cunningham KW, Estruch R (1997) Glucose repression affects ion homeostasis in yeast through the regulation of the stress-activated *ENA1* gene. Mol Microbiol 26:91–98

Arking R, Buck S, Wells RA, Pretzlaff R (1988) Metabolic rates in genetically based long lived strains of *Drosophila*. Exp Gerontol 23:59–76

Arking R, Dudas SP, Baker GT (1993) Genetic and environmental factors regulating the expression of an extended longevity phenotype in a long lived strain of *Drosophila*. Genetica 91:127–142

Austriaco NR, Guarente L (1997) Changes of telomere length cause reciprocal changes in the lifespan of mother cells in *Saccharomyces cerevisiae*. Proc Natl Acad Sci USA 94:9768–9772

Bartholomew JW, Mittwer T (1953) Demonstration of yeast bud scars with the electron microscope. J Bacteriol 65:272–275

Baroni MD, Monti P, Alberghina L (1994) Repression of growth-regulated G1 cyclin expression by cyclic AMP in budding yeast. Nature 371:339–342

Barton AA (1950) Some aspects of cell division in *Saccharomyces cerevisiae*. J Gen Microbiol 4:84–87

Barz WP, Walter P (1999) Two endoplasmic reticulum (ER) membrane proteins that facilitate ER-to-Golgi transport of glycosylphosphatidylinositol-anchored proteins. Mol Biol Cell 10:1043–1059

Berger KH, Yaffe MP (1998) Prohibitin family members interact genetically with mitochondrial inheritance components in *Saccharomyces cerevisiae*. Mol Cell Biol 18:4043–4052

Bodnar AG, Ouellette M, Frolkis M, Holt SE, Chiu CP, Morin GB, Harley CB, Shay JW, Lichtsteiner S, Wright WE (1998) Extension of life-span by introduction of telomerase into normal human cells. Science 279:349–352

Cabib E, Ulane R, Bowers B (1974) A molecular model for morphogenesis: the primary septum of yeast. Curr Top Cell Regul 8:1–32

Chelstowska A, Butow RA (1995) RTG genes in yeast that function in communication between mitochondria and the nucleus are also required for expression of genes encoding peroxisomal proteins. J Biol Chem 270:18141–18146

Chen JB, Sun J, Jazwinski SM (1990) Prolongation of the yeast life span by the v-Ha-*RAS* oncogene. Mol Microbiol 4:2081–2086

Coates PJ, Jamieson DJ, Smart K, Prescott AR, Hall PA (1997) The prohibitin family of mitochondrial proteins regulate replicative lifespan. Curr Biol 7:607–610

Conrad-Webb H, Butow RA (1995) A polymerase switch in the synthesis of rRNA in *Saccharomyces cerevisiae*. Mol Cell Biol 15:2420–2428

Curtsinger JW, Fukui HH, Townsend DR, Vaupel JW (1992) Demography of genotypes: failure of the limited life-span paradigm in *Drosophila melanogaster*. Science 258:461–463

Davidson JF, Whyte B, Bissinger PH, Schiestl RH (1996) Oxidative stress is involved in heat-induced cell death in *Saccharomyces cerevisiae*. Proc Natl Acad Sci USA 93:5116–5121

D'mello NP, Jazwinski SM (1991) Telomere length constancy during aging of *Saccharomyces cerevisiae*. J Bacteriol 173:6709–6713

D'mello NP, Childress AM, Franklin DS, Kale SP, Pinswasdi C, Jazwinski SM (1994) Cloning and characterization of *LAG1*, a longevity-assurance gene in yeast. J Biol Chem 269:15451–15459

Dudas SP, Arking R (1995) A coordinate upregulation of antioxidant gene activities is associated with the delayed onset of senescence in a long-lived strain of *Drosophila*. J Gerontol 50A:B117–B127

Egilmez NK, Jazwinski SM (1989) Evidence for the involvement of a cytoplasmic factor in the aging of the yeast *Saccharomyces cerevisiae*. J Bacteriol 171:37–42

Egilmez NK, Chen JB, Jazwinski SM (1989) Specific alterations in transcript prevalence during the yeast life span. J Biol Chem 264:14312–14317

Engelberg D, Klein C, Martinetto H, Struhl K, Karin M (1994) The UV response involving the Ras signaling pathway and AP-1 transcription factors is conserved between yeast and mammals. Cell 77:381–390

Ewbank JJ, Barnes TM, Lakowski B, Lussier M, Bussey H, Hekimi S (1997) Structural and functional conservation of the *Caenorhabditis elegans* timing gene *clk-1*. Science 275:980–983

Feng J, Funks WD, Wang S-S, Weinrich SL, Avilion AA, Chiu C-P, Adams RR, Chang E, Allsopp RC, Tu J, Le S, West MD, Harley CB, Andrews WH, Greider CW, Villeponteau B (1995) The RNA component of human telomerase. Science 269:1236–1241

Finch CE (1990) Longevity, senescence, and the genome. University of Chicago Press, Chicago

Graves JL, Toolson EC, Jeong C, Vu LN, Rose MR (1992) Dessication, flight, glycogen, and postponed senescence in *Drosophila melanogaster*. Physiol Zool 65:268–286

Hardie DG, Carling D (1997) The AMP-activated protein kinase. Fuel gauge of the mammalian cell? Eur J Biochem 246:259–273

Harley CB, Futcher AB, Greider CW (1990) Telomeres shorten during aging of human fibroblasts. Nature 345:458–460

Hawkins PT, Stephens LR, Piggott JR (1993) Analysis of inositol metabolites produced by *Saccharomyces cerevisiae* in response to glucose stimulation. J Biol Chem 268 : 3374–3383

Heydari AR, Wu B, Takahashi R, Strong R, Richardson A (1993) Expression of heat shock protein 70 is altered by age and diet at the level of transcription. Mol Cell Biol 13 : 2909–2918

Jazwinski SM (1990a) Aging and senescence of the budding yeast *Saccharomyces cerevisiae*. Mol Microbiol 4 : 337–343

Jazwinski SM (1990b) An experimental system for the molecular analysis of the aging process: the budding yeast *Saccharomyces cerevisiae*. J Gerontol 45 : B68–B74

Jazwinski SM (1993) The genetics of aging in the yeast *Saccharomyces cerevisiae*. Genetica 91 : 35–51

Jazwinski SM (1996a) Longevity-assurance genes and mitochondrial DNA alterations: yeast and filamentous fungi. In: Schneider EL, Rowe JW (eds) Handbook of the biology of aging, 4th ed Academic Press, San Diego, pp 39–54

Jazwinski SM (1996b) Longevity, genes, and aging. Science 273 : 54–59

Jazwinski SM (1999a) Molecular mechanisms of yeast longevity. Trends Microbiol 7 : 247–252

Jazwinski SM (1999b) Nonlinearity of the aging process revealed in studies with yeast. In: Bohr VA, Clark BFC, Stevnsner T (eds) Molecular biology of aging. Munksgaard, Copenhagen, pp 35–44

Jazwinski SM, Egilmez NK, Chen JB (1989) Replication control and cellular life span. Exp Gerontol 24 : 423–436

Jazwinski SM, Kim S, Lai C-Y, Benguria A (1998) Epigenetic stratification: the role of individual change in the biological aging process. Exp Gerontol 33 : 571–580

Jiang JC, Kirchman PA, Zagulski M, Hunt J, Jazwinski SM (1998) Homologs of the yeast longevity gene *LAG1* in *Caenorhabditis elegans* and human. Genome Res 8 : 1259–1272

Johnston M (1999) Feasting, fasting and fermenting. Trends Genet 15 : 29–33

Kaibuchi K, Miyajima A, Arai KI, Matsumoto K (1986) Possible involvement of *RAS*-encoded proteins in glucose-induced inositolphospholipid turnover in *Saccharomyces cerevisiae*. Proc Natl Acad Sci USA 83 : 8172–8176

Kale SP, Jazwinski SM (1996) Differential response to UV stress and DNA damage during the yeast replicative life span. Dev Genet 18 : 154–160

Kennedy BK, Austriaco NR, Guarente L (1994) Daughter cells of *Saccharomyces cerevisiae* from old mothers display a reduced life span. J Cell Biol 127 : 1985–1993

Kennedy BK, Austriaco NR, Zhang J, Guarente L (1995) Mutation in the silencing gene *SIR4* can delay aging in *S. cerevisiae*. Cell 80 : 485–496

Kennedy BK, Gotta M, Sinclair DA, Mills K, McNabb DS, Murthy M, Pak SM, Laroche T, Gasser SM, Guarente L (1997) Redistribution of silencing proteins from telomeres to the nucleolus is associated with extension of life span in *S. cerevisiae*. Cell 89 : 381–391

Kenyon C, Chang J, Gensch E, Rudner A, Tabtiang R (1993) A *C. elegans* mutant that lives twice as long as wild type. Nature 366 : 461–464

Khazaeli AA, Tatar M, Pletcher SD, Curtsinger JW (1997) Heat-induced longevity extension in *Drosophila*. I. Heat treatment, mortality, and thermotolerance. J Gerontol 52A : B48–B52

Kim S, Villeponteau B, Jazwinski SM (1996) Effect of replicative age on transcriptional silencing near telomeres in *Saccharomyces cerevisiae*. Biochem Biophys Res Commun 219 : 370–376

Kim S, Kirchman PA, Benguria A, Jazwinski SM (1998) Experimentation with the yeast model. In: Yu BP (ed) Methods in aging research. CRC Press, Boca Raton, pp 191–213

Kimura KD, Tissenbaum HA, Liu Y, Ruvkun G (1997) *daf-2*, an insulin receptor-like gene that regulates longevity and diapause in *Caenorhabditis elegans*. Science 277 : 942–946

King MP, Attardi G (1989) Human cells lacking mtDNA: repopulation with exogenous mito-chondria by complementation. Science 246 : 500–503

Kirchman PA, Kim S, Lai C-Y, Jazwinski SM (1999) Interorganelle signaling is a determinant of longevity in *Saccharomyces cerevisiae*. Genetics 152 : 179–190

Lai E, Prezioso VR, Smith E, Litvin O, Costa RH, Darnell JE (1990) HNF-3A, a hepatocyte-enriched transcription factor of novel structure is regulated transcriptionally. Genes Dev 4:1427–1436

Lakowski B, Hekimi S (1996) Determination of life-span in *Caenorhabditis elegans* by four clock genes. Science 272:1010–1013

Larsen PL (1993) Aging and resistance to oxidative damage in *Caenorhabditis elegans*. Proc Natl Acad Sci USA 90:8905–8909

Larsen PL, Albert PS, Riddle DL (1995) Genes that regulate both development and longevity in *Caenorhabditis elegans*. Genetics 139:1567–1583

Laurenson P, Rine J (1992) Silencers, silencing, and heritable transcriptional states. Microbiol Rev 56:543–560

Lin K, Dorman JB, Rodan A, Kenyon C (1997) *daf-16*: An HNF-3/forkhead family member that can function to double the life-span of *Caenorhabditis elegans*. Science 278:1319–1322

Lin Y-J, Seroude L, Benzer S (1998) Extended life-span and stress resistance in the *Drosophila* mutant *methuselah*. Science 282:943–946

Lindquist SL, Kim G (1996) Heat-shock protein 104 expression is sufficient for thermotolerance in yeast. Proc Natl Acad Sci USA 93:5301–5306

Lithgow GJ, White TM, Melov S, Johnson TE (1995) Thermotolerance and extended life-span conferred by single-gene mutations and induced by thermal stress. Proc Natl Acad Sci USA 92:7540–7544

Loo S, Rine J (1995) Silencing and heritable domains of gene expression. Annu Rev Cell Dev Biol 11:519–548

Luckinbill LS, Arking R, Clare MJ, Cirocco WC, Buck SA (1984) Selection for delayed senescence in *Drosophila melanogaster*. Evolution 38:996–1003

Luckinbill LS, Riha V, Rhine S, Grudzien TA (1989) The role of glucose-6-phosphate dehydrogenase in the evolution of longevity in *Drosophila melanogaster*. Heredity 65:29–38

Luhtala TA, Roecker EB, Pugh T, Feuers RJ, Weindruch R (1994) Dietary restriction attenuates age-related increases in rat skeletal muscle antioxidant enzyme activities. J Gerontol 49:B231–B238

Marchler G, Schuller C, Adam G, Ruis H (1993) A *Saccharomyces cerevisiae* UAS element controlled by protein kinase A activates transcription in response to a variety of stress conditions. EMBO J 12:1997–2003

Martinus RD, Garth GP, Webster TL, Cartwright P, Naylor DJ, Hoj PB, Hoogenraad NJ (1996) Selective induction of mitochondrial chaperones in response to loss of the mitochondrial genome. Eur J Biochem 240:98–103

Masoro E (1995) Dietary restriction. Exp Gerontol 30:291–298

Morishita T, Mitsuzawa H, Nakafuku M, Nakamura S, Hattori S, Anraku Y (1995) Requirement of *Saccharomyces cerevisiae* Ras for completion of mitosis. Science 270:1213–1215

Morris JZ, Tissenbaum HA, Ruvkun G (1996) A phosphatidylinositol-3-OH kinase family member regulates longevity and diapause in *Caenorhabditis elegans*. Nature 382:536–539

Mortimer RK, Johnston JR (1959) Life span of individual yeast cells. Nature 183:1751–1752

Mosch H-U, Roberts RL, Fink GR (1996) Ras2 signals via the *CDC42/STE20/*mitogen-activated protein kinase module to induce filamentous growth in *Saccharomyces cerevisiae*. Proc Natl Acad Sci USA 93:5352–5356

Muller I (1985) Parental age and the life-span of zygotes of *Saccharomyces cerevisiae*. Antonie von Leeuwenhoek J Microbiol Serol 51:1–10

Muller I, Zimmermann M, Becker D, Flomer M (1980) Calendar life span versus budding life span of *Saccharomyces cerevisiae*. Mech Ageing Dev 12:47–52

Murakami S, Johnson TE (1996) A genetic pathway conferring life extension and resistance to UV stress in *Caenorhabditis elegans*. Genetics 143:1207–1218

O'Brien RM, Noisin EL, Suwanichkul A, Yamasaki T, Lucas PC, Wang JC, Powell DR, Granner DK (1995) Hepatic nuclear factor-3 and hormone-regulated expression of the phosphoenolpyruvate carboxykinase and insulin-like growth factor-binding protein 1 genes. Mol Cell Biol 15:1747–1758

Ogg S, Ruvkun G (1998) The *C. elegans* PTEN homolog, DAF-18, acts in the insulin receptor-like metabolic signaling pathway. Mol Cell 2:887–893

Ogg S, Paradis S, Gottlieb S, Patterson GI, Lee L, Tissenbaum HA, Ruvkun G (1997) The forkhead transcription factor DAF-16 transduces insulin-like and metabolic and longevity signals in *C. elegans*. Nature 389:994–999

Orr WC, Sohal RS (1994) Extension of life-span by overexpression of superoxide dismutase and catalase in *Drosophila melanogaster*. Science 263:1128–1130

Pahlavani MA, Haley-Zitlin V, Richardson A (1994) Influence of dietary restriction on gene expression: changes in transcription of specific genes. In: Yu BP (ed) Modulation of the aging process by dietary restriction. CRC Press, Boca Raton, pp 143–156

Paradis S, Ruvkun G (1998) *Caenorhabditis elegans* Akt/PKB transduces insulin receptor-like signals from AGE-1 PI3 kinase to the DAF-16 transcription factor. Genes Dev 12:2488–2498

Parikh VS, Morgan MM, Scott R, Clements LS, Butow RA (1987) The mitochondrial genotype can influence gene expression in yeast. Science 235:576–580

Parkes TL, Elia AJ, Dickinson D, Hilliker AJ, Phillips JP, Boulianne GL (1998) Extension of *Drosophila* lifespan by overexpression of human *SOD1* in motorneurons. Nat Genet 19:171–174

Patton JL, Srinivasan B, Dickson RC, Lester RL (1992) Phenotypes of sphingolipid-dependent strains of *Saccharomyces cerevisiae*. J Bacteriol 174:7180–7184

Pohley H-J (1987) A formal mortality analysis for populations of unicellular organisms (*Saccharomyces cerevisiae*). Mech Ageing Dev 38:231–243

Riha VF, Luckinbill LS (1996) Selection for longevity favors stringent metabolic control in *Drosophila melanogaster*. J Gerontol 51A:B284–B294

Rose MR (1984) Laboratory evolution of postponed senescence in *Drosophila melanogaster*. Evolution 38:1004–1010

Rose MR (1991) Evolutionary biology of aging. Oxford University Press, New York

Rose MR, Vu LN, Parks SU, Graves JL (1992) Selection on stress resistance increases longevity in *Drosophila melanogaster*. Exp Gerontol 27:241–250

Rothermel BA, Thornton JL, Butow RA (1997) Rtg3p, a basic helix-loop-helix/leucine zipper protein that functions in mitochondrial-induced changes in gene expression, contains independent activation domains. J Biol Chem 272:19801–19807

Sabatino F, Masoro EJ, McMahan CA, Kuhn RW (1991) Assessment of the role of the glucocorticoid system in aging processes and in the action of food restriction. J Gerontol 46:B171–B179

Sanchez Y, Taulien J, Borkowich KA, Lindquist SL (1992) HSP104 is required for tolerance to many forms of stress. EMBO J 11:2357–2364

Service PM (1987) Physiological mechanisms of increased stress resistance in *Drosophila melanogaster*. Physiol Zool 60:321–326

Service PM, Hutchinson EW, Mackinley MD, Rose MR (1985) Resistance to environmental stress in *Drosophila melanogaster* selected for postponed senescence. Physiol Zool 58:380–389

Shama S, Kirchman PA, Jiang JC, Jazwinski SM (1998a) Role of *RAS2* in recovery from chronic stress: effect on yeast life span. Exp Cell Res 245:368–378

Shama S, Lai C-Y, Antoniazzi JM, Jiang JC, Jazwinski SM (1998b) Heat stress-induced life span extension in yeast. Exp Cell Res 245:379–388

Sinclair DA, Guarente L (1997) Extrachromosomal rDNA circles – a cause of aging in yeast. Cell 91:1033–1042

Sinclair DA, Mills K, Guarente L (1997) Accelerated aging and nucleolar fragmentation in yeast *sgs1* mutants. Science 277:1313–1316

Sinclair DA, Mills K, Guarente L (1998a) Aging in *Saccharomyces cerevisiae*. Annu Rev Microbiol 52:533–560

Sinclair DA, Mills K, Guarente L (1998b) Molecular mechanisms of yeast aging. Trends Biochem
    Sci 23:131–134
Small WC, Brodeur RD, Sandor A, Fedorova N, Li G, Butow, RA, Srere PA (1995) Enzymatic and
    metabolic studies on retrograde regulation mutants of yeast. Biochemistry 34:5569–5576
Smeal T, Claus J, Kennedy B, Cole F, Guarente L (1996) Loss of transcriptional silencing causes
    sterility in old mother cells of S. cerevisiae. Cell 84:633–642
Smith JR, Pereira-Smith OM (1996) Replicative senescence: implications for in vivo aging and
    tumor suppression. Science 273:63–67
Sun J, Tower J (1999) FLP recombinase-mediated induction of Cu/Zn-superoxide dismutase
    transgene expression can extend the life span of adult Drosophila melanogaster flies. Mol
    Cell Biol 19:216–228
Sun J, Kale SP, Childress AM, Pinswasdi C, Jazwinski SM (1994) Divergent roles of RAS1 and
    RAS2 in yeast longevity. J Biol Chem 269:18638–18645
Tatar M, Khazacli AA, Curtsinger JW (1997) Chaperoning extended life. Nature 390:30
Tatchell K (1993) RAS genes in the budding yeast Saccharomyces cerevisiae. In: Kurjan T (ed)
    Signal transduction: prokaryotic and simple eukaryotic systems. Academic Press, San Diego,
    pp 147–188
Tissenbaum HA, Ruvkun G (1998) An insulin-like signaling pathway affects both longevity and
    reproduction in Caenorhabditis elegans. Genetics 148:703–717
Vanfleteren JR (1993) Oxidative stress and ageing in Caenorhabditis elegans. Biochem
    J 292:605–608
Vanfleteren JR, De Vreese A (1995) The gerontogenes age-1 and daf-2 determine metabolic rate
    potential in aging Caenorhabditis elegans. FASEB J 9:1355–1361
Van Remmen H, Ward W, Sabia RV, Richardson A (1994) Effect of age on gene expression and
    protein degradation. In: Masoro E (ed) The handbook of physiology of aging. Oxford
    University Press, New York, pp 171–234
Vaupel JW, Johnson TE, Lithgow GJ (1994) Rates of mortality in populations of Caenorhabditis
    elegans. Science 266:826
Vélot C, Haviernik P, Lauquin GJ-M (1996) The Saccharomyces cerevisiae RTG2 gene is a
    regulator of aconitase expression under catabolite repression conditions. Genetics 144:
    893–903
Wong A, Boutis P, Hekimi S (1995) Mutations in the clk-1 gene of Caenorhabditis elegans affect
    developmental and behavioral timing. Genetics 139:1247–1259
Xia E, Rao G, Van Remmen H, Heydari AR, Richardson A (1995) Activities of antioxidant
    enzymes in various tissues of male Fischer 344 rats altered by food restriction. J Nutr
    125:195–201
Yan L-J, Levine RL, Sohal RS (1997) Oxidative damage during aging targets mitochondrial
    aconitase. Proc Natl Acad Sci USA 94:11168–11172
Yu CE, Oshima J, Fu YH, Wijsman EM, Hisama F, Alisch R, Matthews S, Nakura J, Miki T,
    Ouais S, Martin GM, Mulligan J, Schellenberg GD (1996) Positional cloning of the Werner's
    syndrome gene. Science 272:258–262

Current Issues Concerning the Role of Oxidative Stress in Aging: A Perspective

Rajindar S. Sohal, Robin J. Mockett, and William C. Orr

Summary: The main tenet of the oxidative stress hypothesis of aging is that accrual of molecular oxidative damage is the principal causal factor in the senescence-related loss of ability to maintain homeostasis. This hypothesis has garnered a considerable amount of supportive correlational evidence, which is now being extended experimentally in transgenic *Drosophila* over-expressing antioxidative defense enzymes. Some of these studies have reported extensions of life span, while others have not. Interpretation of life spans in poikilotherms is complicated by a number of factors, including the interrelationship between metabolic rate and longevity. The life spans of poikilotherms can be extended multifold by reducing the metabolic rate but without affecting the metabolic potential, i.e., the total amount of energy expended during life. A hypometabolic state in poikilotherms also enhances stress resistance and activities of antioxidative enzymes. It is emphasized that extension of life span without simultaneously increasing metabolic potential is of questionable biological significance.

1
Introduction

The phenomenon of senescence or the aging process is characterized by certain shared features among different phylogenetic groups, such as: (1) a progressive loss in the ability to maintain homeostasis, which occurs during the latter part of the life span of an individual animal; (2) variations in the rate of progression of senescence-associated changes and the duration of life spans among individuals and among different species; and (3) the possibility of life-span extension by certain experimental regimens, such as caloric restriction, hypometabolic states in poikilotherms and hibernating mammals, and some single gene mutations or transgenic manipulations in lower organisms such as nematodes and insects. Any unitary mechanism, purporting

Department of Biological Sciences, Southern Methodist University, Dallas, TX 75275, USA

Results and Problems in Cell Differentiation, Vol. 29
Hekimi (Ed.): The Molecular Genetics of Aging
© Springer-Verlag Berlin Heidelberg 2000

to explain the underlying cause(s) of aging, must provide a rational explanation for these various commonalities. It would therefore seem obvious that, in order to clarify the mechanistic underpinnings of the aging process, it will be critical to integrate knowledge obtained from several lines of investigation, using multiple experimental models.

Among all the numerous hypotheses that have been proposed to explain aging, only one, the oxidative stress hypothesis, has attempted to address the aforementioned essential features of the broad phenomenon of aging in animals. Although the validity of this hypothesis has not as yet been thoroughly established, there is presently no specific evidence to invalidate its basic tenets and predictions. The core idea, encapsulated by this hypothesis, is that the rate of aging is dependent upon the rate of accrual of irreversible molecular oxidative damage and that when a certain threshold of cumulative damage is reached, the ability of the homeostatic mechanisms to maintain optimal physiological conditions necessary for survival is lost. A key prediction of this hypothesis is that the rate of aging cannot be retarded, nor can maximum life span be extended, without a corresponding attenuation of oxidative stress/damage.

The oxidative stress hypothesis of aging can be reasonably regarded to be an updated or modified version of the free radical hypothesis of aging, originally advanced by Harman (1956). However, the underlying idea has a longer genesis, which can be clearly traced to Rubner's insights derived from studies in comparative physiology, particularly concerning the phenomena of scaling and physiological time, as discussed elsewhere (Rubner 1908; Schmidt-Nielsen 1984). Rubner pointed out the relative constancy of the energy utilized during adult life, or metabolic potential, among a sample of closely related mammalian species. The inverse relationship between the rate of metabolism and the rate of aging was later formalized as the rate of living theory by Pearl and his associates, and was broadened to encompass multiple phylogenetic groups (Pearl 1928). According to this theory, animals of a given genotype have a fairly fixed metabolic potential, and consequently the length of life is inversely related to the rate at which this energy is consumed, the metabolic rate. The link between rate of living (or metabolic rate) and oxidative stress was subsequently provided by the realization that oxygen utilization necessarily entails the generation of reactive oxygen species (ROS), which can inflict macromolecular damage and alter the redox state of the cells (Chance et al. 1979; Sohal 1986). Moreover, ROS-dependent macromolecular damage is not only detectable in young healthy animals, but the accumulation of such damage increases exponentially during adult life. This led us to propose previously that the level of oxidative stress increases during aging and that, if the rate of accumulation of macromolecular oxidative damage is a major contributor to senescence, then the rate of aging in an animal is an exponential, rather than a linear progressive process (Sohal and Weindruch 1996).

Studies relating to the validity of the oxidative stress hypothesis have passed through several different conceptual and experimental phases. The initial phase dealt with the documentation of age-related alterations in antioxidative defenses and the accumulation of macromolecular oxidative damage (Sohal and Allen 1986). Somewhat later, the rates of mitochondrial ROS generation during aging and in different species were examined (Nohl and Hegner 1978; Farmer and Sohal 1989). This was followed by studies on the effects of caloric restriction, which led to the proposition that life-span extension by this regimen may be due to a decrease in the level of oxidative stress/damage (Sohal et al. 1994b; Sohal and Weindruch 1996). More recently, mutants and transgenic animals have been employed to test the validity of the oxidative stress hypothesis experimentally (Orr and Sohal 1994; Sohal et al. 1995a).

The objective of this chapter is to provide the authors' perspectives concerning the interpretation and significance of the recent studies where mutant and transgenic strategies have been used. An overview of the data obtained in earlier studies of oxidative stress, aging, and longevity provides a broader context for the interpretation of these recent studies. In addition, we have drawn attention to certain issues concerning the relationships between life span, metabolic rate, stress resistance, and antioxidative defenses, which in our opinion are of critical importance in the interpretation of the various results.

2
The Concept of Life Span: A Cautionary Note

Chronological life span is the criterion most commonly used to compare rates of aging. It is widely believed that different genetic strains, subspecies, and species have characteristic chronological life spans. Furthermore, inferences concerning the efficacy of an experimental regimen in affecting the rate of aging are quite often based on its effect on duration of life. While this concept of a characteristic or typical life span is quite rational when applied to homeotherms with relatively stable metabolic rates, it is inappropriate and potentially misleading in the case of poikilotherms, which have variable rates of metabolism. The life spans of poikilotherms such as insects and nematodes, which are the most widely used models for aging research among lower animals, are profoundly dependent on metabolic rate. Consequently, no reliable inferences can be drawn about aging in homeotherms solely on the basis of chronological life-span variations in poikilotherms. Such inferences can only be drawn if the rate of metabolism is determined simultaneously and the metabolic potential is shown to have increased. For example, by decreasing the ambient temperature from 30 to 10°C, the life span of

Drosophila melanogaster can be extended 8.9-fold (Loeb and Northrop 1917); however, the metabolic potential tends to remain unaltered at different temperatures (Miquel et al. 1976; McArthur and Sohal 1982; Sohal 1982). In other words, at different ambient temperatures, the total amount of oxygen consumed during the course of a life span remains relatively constant. Other experimental regimens which affect the rate of metabolism also influence life span. For instance, prevention of flying activity in houseflies results in a two- to three-fold increase in life span (Ragland and Sohal 1973), but no significant alteration in metabolic potential (Sohal 1982).

The effect of metabolic rate in poikilotherms is so profound that even potentially deleterious treatments can extend life span, provided that metabolic rate is decreased and the severity of damage is below a lethal threshold. For instance, it has been noted in a variety of animal species that exposure to X-irradiation extends life span. In the case of male houseflies, life span was extended 20% following exposure to 15 kr irradiation, but the metabolic potential was decreased by 7% (Allen and Sohal 1982). Thus, relying solely on the criterion of chronological life span would mask the otherwise deleterious effects of X-irradiation, as revealed by the decrease in amount of metabolic energy expended during life, i.e., the metabolic potential. Similarly, treatments such as the administration of ethidium bromide (Fleming et al. 1981), which intercalates DNA, and diethyldithiocarbamate, which inactivates Cu-Zn superoxide dismutase (SOD), can extend life span by up to 45% in insects, but such treatments invariably decrease the metabolic rate (Sohal et al. 1984b). The extensions in chronological life span appear to be due to the beneficial effects of entry into a hypometabolic state.

It is thus quite obvious that a rational interpretation of the results of any experimental regimen, purporting to affect the rate of aging, especially in poikilotherms, should take into consideration whether or not the metabolic potential has been affected. An extension of chronological life span without any gain in metabolic potential is virtually meaningless from a biological perspective, especially in terms of a beneficial manipulation of the aging process.

Unfortunately, many reports of life-span extension in nematodes and insects, based on single gene mutations, transgenic manipulations, or dietary supplementation with antioxidants, have not been accompanied by careful documentation of metabolic potential. Indeed, mutations that produce biochemical defects, resulting in lethargic animals with reduced metabolic rate but increased chronological life spans, have sometimes been claimed to identify gerontogenes which govern the aging process (Johnson et al. 1999). An alternative interpretation is that mutant and transgenic manipulations that produce sluggish phenotypes with lowered metabolic rates, but extended longevity, simply provide genetic confirmation of the importance of metabolic rate in the determination of life span, and implicitly of the rate of living theory (Van Voorhies and Ward 1999).

3
Metabolic Rate, Stress Resistance and Antioxidative Defenses

A frequently reported feature of most of the mutants and transgenics exhibiting extended longevity is the enhancement of resistance to stresses such as paraquat administration, desiccation, or starvation. This has been interpreted to indicate that the relevant genes encode enzymes involved in stress resistance (Jazwinski 1996; Parsons 1996; Johnson et al. 1999). An increase in the activity of Cu-Zn SOD and catalase in the *age-1* mutant of *Caenorhabditis elegans* has been cited as an indication of the possible mechanistic basis for enhanced resistance to paraquat toxicity and extension of life span (Johnson et al. 1999). An alternative interpretation, based on a vast body of literature, is that any stressful treatment that causes a decrease in the rate of metabolism also leads to an increase in resistance (or decrease in responsiveness) to other additional stresses. Furthermore, the decrease in metabolic rate under stressful conditions almost invariably causes an increase in activities of antioxidative enzymes (reviewed by Hermes-Lima et al. 1998). This is regarded as a fundamental adaptive response to stresses that elicit the hypometabolic state. Thus, increased resistance to stress and elevation in antioxidative defenses are markers of a hypometabolic state, but not for gerontogenes as proposed by some authors.

This interpretation is based on the following evidence. (1) A decrease in body temperature and consequently in metabolic rate provides protection against oxygen toxicity (Campbell 1936). (2) Virtually all of the biological effects of irradiation are less pronounced under hypometabolic conditions such as hypoxia and anoxia. Furthermore, insects such as the beetle, *Tribolium*, or *Drosophila* become relatively more resistant to oxygen toxicity after induction of a hypometabolic state through sublethal irradiation (Lamb and McDonald 1973; Lee and Ducoff 1984). The irradiated flies are also more resistant to dry heat stress than the controls (Lamb and McDonald 1973). (3) Food and/or water deprivation has a protective effect against additional stresses in various living systems. For instance, such deprivation protects rats against hyperoxia-induced seizures in a dose-dependent manner (Bitterman et al. 1997). Fasting mice survive longer under irradiation and oxygen poisoning (Gilbert et al. 1955). Moreover, bacteria such as *E. coli* or *Vibrio parahaemolyticus* are more resistant to thermal, osmotic, or H_2O_2-induced stress under starved versus normal conditions (Koga and Takumi 1995). (4) Estivation (dormancy) in pulmonate land snails such as *Helix* sp. under dry conditions results in depression of metabolic rate, typically to ≤30% of the resting rate in active organisms, and an elevation in the activity of antioxidative enzymes such as SOD and catalase (Hermes-Lima et al. 1998). (5) There is a vast amount of evidence that many stressful conditions, such as freezing, hypoxia, anoxia, desiccation, and others that induce dormancy or estivation in poikilotherms, also reduce metabolic rate and elevate the activities of antioxidative enzymes. This would at first appear paradoxical since, under such hypometabolic conditions, rates of ROS generation are expected to decline. Nevertheless, this elevation in antioxidative defenses is thought to constitute a fundamental adaptive strategy by animals to endure environmental stresses that cause wide variations in the availability of oxygen (Hermes-Lima et al. 1998).

The information summarized above is consistent with the broad generalization that genetic manipulations or experimental regimens, that are manifestly deleterious, may extend chronological life span by a reduction in metabolic rate, and an overall lowering of metabolic potential. A consequence of the decreased metabolic rate is an increase in resistance to various stresses accompanied by an elevation in antioxidative enzyme activities. The crux of this argument is that the hypometabolic state is the primary cause for enhanced tolerance to stress, increased chronological life span, and elevated activities of antioxidative enzymes. We therefore suggest that these issues be taken into account in interpreting the results of any study purporting to manipulate the rate of aging in lower organisms. Speculations that gerontogenes encode some novel and unknown pathway of stress resistance (Johnson et al. 1999; Lin et al. 1998) may be totally erroneous, and may unnecessarily confuse efforts to elucidate causal factors associated with aging.

4
Current Evidential Status of the Oxidative Stress Hypothesis of Aging

Current versions of the oxidative stress hypothesis of aging state, essentially, that aging is due, at least in part, to an imbalance between prooxidant production and antioxidant defenses/repair processes. This oxidative imbalance is believed to result in accumulation of molecular damage, which underlies age-associated physiological attrition. If oxidative stress is a significant causal factor in aging, then the following predictions should be confirmed (Sohal and Weindruch 1996). First, among members of a given species, the older animals should have higher steady-state levels of damage byproducts, reflecting the net effect of the imbalance between prooxidant production and antioxidative defenses/damage repair. If oxidative damage byproducts accumulate at a constant rate, there may be no age-related change in the individual components of the oxidative network, whereas if damage accumulates at an accelerating rate, there should be age-related increases in prooxidant production and/or decreases in defense or repair capability. Second, in comparisons within and among species, those with longer life spans should have lower levels of oxidative damage byproducts at a given age. They should have lower levels of prooxidant production, or higher levels of defense or repair capability, and slower age-related changes in these parameters. Third, experimental manipulation of the level of oxidative stress should have corresponding effects on longevity. For example, life spans should be decreased by exposure to oxidizing environments, such as 100% oxygen or X-irradiation, or by dietary supplements which increase prooxidant production, such as paraquat. Mutants with lower levels of antioxidative defenses should also have shorter life spans. Conversely, bolstering antioxidant defenses by dietary supplementation or creation of transgenic animals should increase longevity.

The current experimental status of these predictions, in our assessment, is summarized in Table 1. Since mitochondrial respiration is an important source of ROS, with approximately 2% of the oxygen utilized undergoing successive univalent reductions (Chance et al. 1979) to superoxide anion (O_2^{-}) and hydrogen peroxide (H_2O_2), which may then undergo further reduction to the highly reactive hydroxyl free radical ($^{.}OH$), mitochondrial ROS (O_2^{-} and H_2O_2) have been used as indices of prooxidant production. Production of these species has been shown to rise as a function of age in multiple organs of mammalian species (Nohl and Hegner 1978; Sohal et al. 1990b), and in dipteran insects (Sohal and Sohal 1991). The levels of prooxidant production also vary inversely with the maximum life-span potential (MLSP) of numerous species, both within and between different animal phyla (Sohal et al. 1989, 1990d, 1995b; Ku et al. 1993). As expected, experimental elevation of prooxidant production by hyperoxia, X-irradiation, or dietary supplementation with various oxidants decreases longevity (Allen et al. 1984; Sohal 1988; Sohal and Dubey 1994). Experimental decreases in prooxidant production, which would directly and convincingly test the hypothesis, have been less easy to achieve.

The antioxidant component of the oxidative balance has been studied most extensively, not only because it comprises a network of well-characterized enzymatic and small molecular weight species, but also because this compo-

Table 1. Current evidential status of the oxidative stress hypothesis of aging[a]

	Prooxidant production	Antioxidant defenses	Repair processes	Oxidative damage
Age-related changes	+	+/−	+/−/?	+
Correlation with life-span variation	+	+/−	+/?	+
Effects of experimental modulation	+	+/−	?	+
Net evidence	+	+/−	+/?	+

[a] + indicates that, in the authors' view, the balance of evidence in a particular category supports the oxidative stress hypothesis of aging; − indicates that the evidence opposes or fails to support the predictions of the hypothesis; ? indicates that testing of the hypothesis has not been sufficiently comprehensive to determine whether or not its predictions will be confirmed. Levels of oxidative damage are considered to reflect the net effect of variations in prooxidant production, antioxidant defenses, and repair processes.

nent seems most promising for augmentation of human life, if the hypothesis is true. Consequently, a very large number of studies have been conducted on the effects of antioxidant supplementation, and many of these studies have shown extension of life span (Comfort et al. 1971; Harman 1978; Miquel and Economos 1979; Ruddle et al. 1988). However, this improvement is sometimes quite limited (Miquel et al. 1982); it frequently affects only mean rather than maximum life span (Kohn 1971; Harman 1981), suggesting the supplements may correct nutritional deficiencies rather than affecting aging per se (or that sample sizes are too small to demonstrate changes in maximum life span); and some studies show no effect or even a harmful effect of antioxidant supplementation (Massie et al. 1976; Herbert 1994; The Alpha-Tocopherol, Beta-Carotene Cancer Prevention Study Group 1994). Additionally, bolstering enzymatic antioxidants by the transgenic approach does not consistently result in increased longevity (Seto et al. 1990; Orr and Sohal 1993). In comparisons between mammalian species with differences in MLSP, there is no obvious connection between overall levels of antioxidant defenses and longevity (Sohal et al. 1990c), although there is some evidence for such a connection in *Drosophila* (Bartosz et al. 1979; Fleming et al. 1987). In comparisons between young and old animals, the levels of individual antioxidants rise, fall, or remain unchanged, in a tissue- and species-specific pattern (Sohal et al. 1984a, 1990b). Changes in redox ratios suggest a net decrease in antioxidative capability with advancing age, but the effect is relatively small (Sohal et al. 1990a).

The repair of oxidative damage is a third important component affecting the oxidative balance, which until recently has received comparatively little attention in studies of aging. Nonetheless, a number of repair processes for DNA (Breimer 1983; Hollstein et al. 1984), lipid and protein (Pacifici and Davies 1991; Grune et al. 1997) oxidative damage have been identified. Studies in rat hepatocytes have shown a substantial age-related decrease in alkaline protease activity (Starke-Reed and Oliver 1989), but the magnitude of decrease in alkaline protease activity appears to vary widely between tissues, and is not observed in houseflies (Agarwal and Sohal 1994a). Evidence for age-related decreases in repair of DNA oxidative damage is not conclusive, with some studies supporting the idea and others contradicting it (Ishikawa and Sakurai 1986; Bohr and Anson 1995; Hirano et al. 1996), but the repair of such damage was found to be proportional to the logarithm of the life span in seven mammalian species (Hart and Setlow 1974). Similarly, the levels of excreted DNA oxidation repair products were higher in shorter-lived mammalian species (Adelman et al. 1988). In houseflies, repair capability was found to be lower in shorter-lived members of a single population, and higher in the longer-lived members (Newton et al. 1989). In *Drosophila*, a mutant deficient in DNA repair was found to have decreased longevity and mating activity (Miquel et al. 1983). A more systematic study of age-associated changes in repair activities in different species is needed to determine the role of repair in the oxidative imbalance associated with aging.

Oxidative damage represents the net effect of prooxidant, antioxidant, and repair components of the oxidative balance. Thus, while unexpected findings with antioxidants or other components can be reconciled with the oxidative stress hypothesis, detection of oxidative damage accumulation is a necessary requirement for verification of the hypothesis. There have been widespread findings of oxidative damage accumulation during aging of all animal species investigated to date. Besides increases in DNA (Agarwal and Sohal 1994b) and protein oxidation adducts (Oudes et al. 1998), and lipid peroxidation byproducts (Sohal et al. 1985b; Yu 1996), there is a steady increase in lipofuscin, an autofluorescent age-pigment which is thought to consist of indigestible, cross-linked lipid and protein oxidation byproducts (Sohal and Wolfe 1986; Sohal and Brunk 1990). There are also higher levels of oxidation byproducts in comparatively short-lived animal species (Sohal et al. 1993), and in animals with shorter life expectancy within a single population (Sohal and Dubey 1994). Finally, experimental measures which increase oxidative stress and decrease longevity, such as exposure to 100% oxygen or X-irradiation, cause rapid accumulation of oxidative damage (Agarwal and Sohal 1994c; Sohal and Dubey 1994). Measures which prolong life, such as caloric restriction in mammals (Sohal et al. 1994a,b) or reduced physical activity in flies (Sohal and Dubey 1994), retard the accrual of damage byproducts.

The measurement of accumulating oxidative damage to all of the major classes of biological molecules is of central importance to verification of the oxidative stress hypothesis of aging. However, oxidative damage cannot be experimentally manipulated in a direct manner. Instead, changes in oxidative damage can only be produced indirectly, as a consequence of experimental variations in the components of the oxidative balance. Since the means to reduce prooxidant production are presently lacking, and information about the importance of damage repair in the oxidative balance has been comparatively meager, antioxidant elevation has been the main focus of research designed to decrease oxidative stress and slow aging. One important focus of this work has been the creation of transgenic fruit flies, *Drosophila melanogaster*, by the P element-mediated transformation technique.

5
Longevity Studies in Transgenic *Drosophila*

The majority of studies involving antioxidative enzyme overexpression in transgenic *Drosophila* have concentrated on the cytosolic enzyme, Cu-Zn superoxide dismutase (SOD). Seto et al. (1990) reported that homozygous lines overexpressing *Drosophila* Cu-Zn SOD by 30–70% had essentially the same life span as the isogenic wild type Oregon R line. Examination of the survival curves from this study suggests that the median and maximum

longevity were increased about 10% at 29°C in the SOD overexpressors, but
there was no increase in resistance to the superoxide-generating herbicide,
paraquat. In a second study, heterozygous lines overexpressing *Drosophila*
Cu-Zn SOD by 32–42%, under the control of native promoter and regulatory
sequences, were found to have no increase in maximum longevity or resistance to paraquat (Orr and Sohal 1993). Only one line had a small improvement in mean survival time under normal and hyperoxic conditions
(20% oxygen and 100% oxygen, respectively).

There have been several studies of the effects of Cu-Zn SOD overexpression using ectopic promoters and heterologous SOD genes. In the first of
these studies, bovine Cu-Zn SOD was expressed in transgenic flies under the
control of the *Drosophila* actin 5C gene promoter (Reveillaud et al. 1991).
Elevation of total SOD activity by an average of 32.5% in these lines was
associated with increases of up to 17% in mean life span. There was no
increase in maximum survival time, but an increased resistance to paraquat
was noted. Expression of the bovine SOD alone, in a null strain with no
Drosophila Cu-Zn SOD activity and a drastically curtailed life span, resulted
in partial rescue (up to 30%) of normal survival times (Reveillaud et al.
1994). Much more recently, Parkes et al. (1998) reported that targeted expression of human Cu-Zn SOD in *Drosophila* motor neurons, using the yeast
GAL4/UAS expression system, extended mean and maximum survival times
by up to 40 and 30%, respectively. In a null background, this targeted expression restored survival times from 5% to more than 60% of the wild-type
values. Finally, inducible overexpression of *Drosophila* Cu-Zn SOD controlled
by the actin 5C promoter was reported to increase the mean life span by up to
48%, depending on the genetic background of the flies (Sun and Tower 1999).
Expression of the transgene was induced in young adult flies by expression of
the yeast FLP recombinase in a binary transgenic system (see Sect. 6.3).

Additional studies of antioxidant overexpression in transgenic flies have
involved catalase and Cu-Zn SOD/catalase "double dose" transgenics. Enhancement of catalase activity by 70–80% produced a 10% increase in mean
life span in one transgenic line (Orr and Sohal 1992). There was no increase
in a second line, and no effect on maximum longevity or resistance to
hyperoxia or paraquat. Resistance to H_2O_2, the substrate for catalase, was
significantly increased in both transgenic lines. Using the FLP recombinase
expression system, *Drosophila* catalase overexpression increased resistance to
H_2O_2, but had no beneficial effect on survival in either of two genetic
backgrounds (Sun and Tower 1999).

The transgenic strategy was also used to generate flies overexpressing two
antioxidative enzymes simultaneously from separate transgenes. Tandem
overexpression of Cu-Zn SOD and catalase in an isogenic background extended median and maximum life spans up to 34% (Orr and Sohal 1994). The
increase in longevity was accompanied by slower accrual of protein and DNA
oxidative damage, and increased resistance to experimental oxidative stress

(Sohal et al. 1995a). The same enzymes were also overexpressed simultaneously using the FLP recombinase system. Several lines were found to have either increased or decreased life spans, but the percent increase was less in the presence of both transgenes than with the Cu-Zn SOD transgene alone.

Finally, it is perhaps of some relevance to this survey to include trangenic studies involving hsp70. Transient overexpression of hsp70 in transgenic flies increased age-specific survival rates in young adults (Tatar et al. 1997). Use of an hsp70-β-galactosidase fusion reporter transgene showed tissue-specific, age-associated induction without heat shock (Wheeler et al. 1995). The reporter transgene was also induced in mutant flies lacking Cu-Zn SOD or catalase, and age-associated induction was accelerated in catalase hypomorphs with decreased life span. Collectively, these results suggest a link between oxidative stress, hsp70 expression, and, by extension, the improved survival of hsp70 overexpressors.

6
Hazards of Life-Span Analysis in *Drosophila*

The foregoing synopsis of research findings from transgenic flies implicitly assumes a close correspondence between length of life and rate of aging, while omitting discussion of potential hazards in the interpretation of survival data. In reality, an increase in the length of life cannot be equated automatically with a decrease in the rate of aging. Neglecting issues related to the statistical analysis of the significance of life-span extension, which have been discussed extensively elsewhere (Parmar and Machin 1995; Manton et al. 1991), the potential problems can be grouped into three categories. First, compensatory changes may occur within the animal, in response to introduction of a transgene (or dietary supplementation with a chemical). Second, effects of a transgene in a particular genetic background may reflect idiosyncracies of the specific strain used. Such effects may be masked or spuriously enhanced by any genetic differences between animals used in an experiment. Third, the use of ectopic regulatory sequences, foreign genes, and inducible expression systems, often in an attempt to avoid genetic variation, introduces a number of variables in addition to the increased enzyme activity as potential causes for any changes in longevity.

6.1
Compensation

Either biochemical or physiological compensatory changes may take place in response to increases in a particular antioxidant. The most likely biochemical response, a decreased activity of one antioxidant in response to increases in another, results in no net enhancement of the antioxidative network. Such biochemical compensation has been most obvious in mutant animals with severe deficiencies of a single antioxidant, in which other defenses were naturally elevated (Orr et al. 1992). Compensatory decreases in endogenous antioxidants have been noted in response to dietary supplementation with exogenous antioxidants (Sohal et al. 1985a). There have been no reports of obvious decreases in other defenses in response to antioxidant elevation in transgenic flies, possibly because the total antioxidant capability is normally in excess of that required for survival, but in transfected V79 Chinese hamster cells, a 2.2- to 3.5-fold elevation of Cu-Zn SOD activity was associated with significant decreases in other antioxidative enzyme activities (Teixeira et al. 1998). Thus, each time a single antioxidant is bolstered, the levels of the others must be tested to ensure that there is actually a net increase in antioxidative capability. Regrettably, in several studies of transgenic flies, the levels of other antioxidants have not been reported. Even when the main components of the antioxidative network are measured, others are not examined. Consequently, the basis for interpretation of these studies as supporting or refuting the oxidative stress hypothesis is diminished. Unfortunately, there is no single measure of net antioxidative capability. Although ratios of redox couples may be used for this purpose, their levels are also regulated in response to other needs of the cell (Hwang et al. 1992).

In comparison with biochemical compensation, physiological compensation is arguably an even more widespread and fundamental response to experimental manipulation of poikilothermal animals, including *Drosophila* (Sohal and Allen 1986). As discussed above, such compensation reflects the ability of poikilotherms to effectively decrease their rate of living by slowing their metabolic rate. Thus, in the case of transgenic flies with increased antioxidative enzyme activity, but unchanged longevity, decreased oxygen consumption would imply that the transgene was harmful, while increased oxygen consumption would signify a beneficial effect, allowing the flies to maintain a higher metabolic rate without dying sooner. A particularly crucial experiment with flies overexpressing Cu-Zn SOD and catalase showed that the metabolic potential was increased by about 30%, corresponding closely to the increase in longevity (Sohal et al. 1995a). It is unfortunate that in only one other study (Parkes et al. 1998) is any effort made to measure metabolic potential, and that in studies of long-lived mutants this issue has been largely ignored. Thus, based on a recent report of a genetic mutant with increased longevity and stress resistance (Lin et al. 1998), some authors have claimed that aging in all animals is almost certainly controlled by a genetic program (Pennisi 1998). Since measurements of the metabolic potential of the mutant are not yet available, not only are such generalizations premature, in fact the relevance of the mutation to aging cannot be assessed.

6.2
Genetic Effects

Genetic variation between animal lines and strains is a crucial variable in longevity studies, since uncontrolled genetic differences can produce differences in survivorship unrelated to the effects of the transgene itself. Introduction of a transgene by microinjection or transgene mobilization immediately creates genetic variation, since the transgene inserts at a different position in each line, potentially perturbing gene expression at neighbouring loci. The effect varies among lines with different insertion sites, and in each case differs from the parental strain, which has no transgenes at any of the insertion positions. Such position effects may also affect the level of expression of the transgene, and the interaction with neighbouring loci may influence the fitness and life span of the animal. Furthermore, each line is derived from a single injected embryo, which may have genetic differences with respect to other embryos from a heterogeneous parental population.

In view of the genetic variation that characterizes this standard transgenic approach, the most effective control is to increase the number of independent lines tested. As the number of lines increases, any net consequence that may obtain due to position effects diminishes. This possibility is further diminished by selecting only those lines for which adults homozygous for the transgene (or transgenes) can be recovered and used to generate stable stocks. These lines are then backcrossed to the parental line to generate heterozygous progeny for experimentation. In practice, this approach may yield sets of more than 20 lines, in which none differs from the others in longevity due to obvious position effects, although variations between cohorts may be triggered by any variation in environmental conditions (unpubl. observ.). A further advantage of this approach is that stocks may be outcrossed to confirm observations in other genetic backgrounds.

This ability to examine survivorship of transgenic flies in multiple genetic backgrounds takes on added significance when apparent trends in *Drosophila* longevity studies are considered. In the various studies with transgenic flies, and in studies involving dietary supplementation, it is notable that life-span extension is generally somewhat greater when the control strain has a short life span, and lesser when the control life span is longer (Fig. 1). This phenomenon may result because of strain-specific weaknesses, distinct from the broader phenomenon of aging, which are compensated for (rescued) by transgene products or dietary supplements, creating an artificial impression that aging itself has been slowed. This interpretation is consistent with the finding of Kohn (1971), who observed that dietary antioxidant supplementation of mice increased life span only when the survival of control animals was suboptimal.

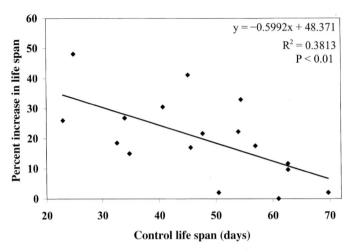

Fig. 1. Correlation between life-span extension in *Drosophila melanogaster* and length of the control life span. Data are taken from various studies of antioxidant elevation by dietary supplementation or the transgenic approach. Studies involving mutations, transgenes, or supplements were excluded from the analysis if life-span extension could not be attributed to an antioxidative effect. Results represent the maximum percent increase in either mean or median life span obtained at 25°C in each study. The correlation shows a significant negative relationship between the magnitude of life-span extension and length of the control life span. However, the wide scatter of the data indicates that other factors have a major influence on life-span extension, including differences in effectiveness of the various antioxidative strategies which were examined, and possibly differences in experimental conditions and *Drosophila* strains employed in various laboratories. (Data were obtained from the following sources. For *transgenic studies*: Seto et al. 1990; Reveillaud et al. 1991; Orr and Sohal 1992, 1993, 1994; Parkes et al. 1998; Mockett et al. 1999; Sun and Tower 1999. For *supplementation studies*: Massie et al. 1976, 1993; Massie and Williams 1979; Miquel et al. 1982; Izmaylov and Obukhova 1996, 1999; Anisimov et al. 1997, 1998; Brack et al. 1997; Yong-Xing et al. 1997)

Considering the concentration of research effort on Cu-Zn SOD overexpression, the specific question which should be readily answerable based on these studies is whether overexpression of this enzyme lengthens life span in *Drosophila*. The existing studies conclude that overexpression of Cu-Zn SOD has either a large beneficial effect (up to 48%), a small but significant beneficial effect, or no effect on life span. Both theoretical considerations (McCord 1995) and experimental evidence (Omar et al. 1990) suggest that there may be an optimum concentration of this enzyme, while either higher or lower levels lead to increased oxidative stress. Thus, the conflicting results may reflect strain-specific variations in the optimum level of SOD activity, or in the evolved level of SOD activity in relation to the optimum. The tendency of short-lived strains to gain a greater life-span extension, following enhancement of their antioxidative defense capabilities, may thus reflect higher levels of oxidative stress or suboptimal endogenous defenses, which could, in turn, account for the shorter life spans of these strains.

It is also possible that variations in handling or diet could contribute to differences in longevity, in which case the beneficial effects of transgenes in shorter-lived flies may be due to compensation for environmental stresses rather than aging per se. Of particular significance in this respect, many investigators report that flies are transferred to fresh food containers only every 2 or 3 days, whereas in our experience fresh containers must be provided to older flies (>20–40 days, varying by cohort) every 24 h to observe a full-length life span. Thus, notwithstanding the hazards of comparing the absolute length of life obtained in different laboratories, involving differences of both genotype and environment, there are valid reasons to consider the absolute length of life at a given temperature when assessing the connection between rates of aging and life-span extension.

6.3
Inducible Expression Systems

Although the problems of genetic variation may be minimized by using multiple independent, outcrossed transgenic lines, inducible expression systems have the unique advantage of controlling fully for genetic variation. In these systems, both control and experimental flies contain the same transgenes in the same positions, but the controls do not express the anti-oxidative transgene because the required inducing stimulus is not applied. Use of heat-shock gene promoters allows induction by application of a heat pulse. Recently, two binary transgenic systems have been developed to provide controlled transgene expression. In the tet-on system, one transgene constitutively expresses the reverse tetracycline repressor (rtR) fused to the transcriptional activation domain of herpes virus protein VP16 (Bieschke et al. 1998). The second, regulated transgene is under the control of the Adh or hsp70 core promoter, fused to seven tetracycline operator (tetO) sequences. Upon feeding with doxycycline, a tetracycline derivative, the rtR domain binds to the operator and expression is induced. In the FLP recombinase system, expression of the recombinase from one transgene is driven by the hsp70 promoter following a heat pulse (Sun and Tower 1999). The second transgene consists of a coding sequence, separated from the constitutive actin 5C promoter by a transcriptional stop sequence, which is excised upon expression of FLP recombinase from the first transgene. Besides full control for genetic variation, these systems offer two additional advantages: (1) overexpression of an antioxidative enzyme may be initiated at any time during development or adult life; and (2) such systems could be exploited to initiate patterns of spatial expression which differ from that of the native gene product, as described previously for expression of human Cu-Zn SOD using the yeast GAL4/UAS expression system (Parkes et al. 1998).

However, the use of foreign genes and abnormal patterns of spatial and temporal expression also introduces additional variables which may complicate the interpretation of results of such studies.

Notwithstanding their advantages, the inducible expression systems are subject to several disadvantages. For instance, in the FLP recombinase study involving catalase and Cu-Zn SOD, expression of catalase was often substantially increased in control animals even in the absence of the heat pulse. The most likely interpretation of this finding is that overexpression of antioxidative transgenes can occur in some tissues of control flies, suggesting some expression of FLP recombinase in the absence of a heat pulse, although this effect was not observed with the SOD transgene. It is also at least theoretically possible with this system that some cells in the experimental animals would not express a functional transgene product, unless the heat pulse was sufficient to ensure universal expression of the recombinase in every cell type (although preliminary studies with a *lacZ* reporter construct demonstrated that the majority of the fly DNA underwent the recombinase-mediated excision reaction, with increased gene expression in all tissues). This kind of uncertainty about spatial variation in transgene overexpression seems to hold at least the same potential as genetic variation (position effects) to affect the interpretation of the experiment, and to be harder to monitor.

Direct effects of the inducing system may further complicate interpretation of the results obtained with such systems. For instance, in the FLP recombinase system, the heat pulse has a direct negative effect on survival, but this effect varies in magnitude between different cohorts of the same genotype. In comparisons with control flies which are not exposed to heat, the extent to which life-span elevation in experimental flies is affected owing to their more stressful life history is different in each experiment. Thus, the effects of the inducing system and the second transgene product are difficult to resolve from one another.

An additional problem with the study employing FLP recombinase is the use of strains bearing dominant mutations, such as *Stubble*, in order to test the effects of transgenes in different genetic backgrounds. While the use of different backgrounds provides an important control for strain-specific effects of a particular transgene, the presence of dominant mutations with obvious phenotypic effects may impose a life-shortening stress on the animal. Accordingly, the transgene product may extend life span in such backgrounds by counteracting this stress, which may not be comparable with the normal aging process in wild-type flies. To avoid this problem, the effects of transgenes should be tested in different wild-type backgrounds, such as Oregon-R and Canton-S. In conclusion, in spite of their obvious advantages, the uncertainties introduced by the inducible expression systems are potentially greater and less readily quantitated than those introduced by simpler transgenic systems.

7
Conclusions

The oxidative stress hypothesis of aging is supported in a fairly consistent manner by correlative evidence linking prooxidant production, oxidative damage, aging, and life-span variations. The evidence obtained from studies of antioxidant defenses is less consistently supportive of the idea, although it does not undermine it, while evidence concerning repair processes is not sufficiently comprehensive to support or refute the hypothesis. Experimental manipulations designed to test the hypothesis have generally involved life-span reduction or attempts to extend the life span by augmenting antioxidant defenses. The existing evidence for life-span extension resulting from elevation of antioxidant defenses in transgenic *Drosophila* is inconclusive. In several studies, this approach resulted in considerable increases in life span, which were associated in some instances with comparable decreases in the accrual of oxidative damage, and with increases in metabolic potential. Conversely, in several studies, overexpression of antioxidative enzymes produced no effect on either oxidative damage or longevity. Insofar as the changes in longevity were correlated with changes in oxidative stress and damage, the results obtained with transgenic flies were consistent with the oxidative stress hypothesis of aging. However, the limited effects of antioxidant transgenes in many cases were also reminiscent of earlier data from dietary supplementation experiments and comparisons between and within species during aging, many of which suggested that antioxidant defenses are not the limiting component of the oxidative balance.

Any life-span extension (or absence thereof) in poikilotherms should be understood in terms of possible biochemical and physiological compensation before extrapolating the results to higher animals. In many cases, the possibility of compensatory effects has not been examined in transgenic flies, but compensation has frequently been observed in various manipulations of poikilothermal animals, involving alteration of either the metabolic rate or the levels of endogenous antioxidants. Thus, the potential of antioxidant enzyme overexpression to provide a critical test of the oxidative stress hypothesis of aging has not been exploited fully. To realize the potential of the transgenic system, the flies must be characterized not only for their survival times, but also for levels of other antioxidants, oxygen consumption, and physical activity. In addition, transgenic flies should be generated which overexpress antioxidant enzymes or combinations of enzymes in addition to Cu-Zn SOD and catalase.

Finally, if studies of this kind yield limited increases in longevity, the transgenic model system may be used subsequently to enhance repair capability or even to decrease prooxidant production. Although the latter undertaking would be very ambitious, realistic methods can be envisioned in which exogenous antioxidative defenses could be targeted to major sites of prooxidant production, such as the mitochondria. From the perspective of extramitochondrial targets of mitochondrial oxidants, the effects of such strategies would be more akin to decreased prooxidant production than to elevated defenses. Given the tighter link between life expectancy and pro-oxidant production, versus antioxidant defenses, correspondingly more spectacular life-span extension may be predicted to follow from success with this approach.

Acknowledgments. The research program of the authors is supported by grants from the National Institutes of Health – National Institute on Aging.

References

Adelman R, Saul RL, Ames BN (1988) Oxidative damage to DNA: relation to species metabolic rate and life span. Proc Natl Acad Sci USA 85:2706–2708

Agarwal S, Sohal RS (1994a) Aging and proteolysis of oxidized proteins. Arch Biochem Biophys 309:24–28

Agarwal S, Sohal RS (1994b) DNA oxidative damage and life expectancy in houseflies. Proc Natl Acad Sci USA 91:12332–12335

Agarwal S, Sohal RS (1994c) Aging and protein oxidative damage. Mech Ageing Dev 75:11–19

Allen RG, Sohal RS (1982) Life-lengthening effects of γ-radiation on the adult housefly, *Musca domestica*. Mech Ageing Dev 20:369–375

Allen RG, Farmer KJ, Newton RK, Sohal RS (1984) Effects of paraquat administration on longevity, oxygen consumption, lipid peroxidation, superoxide dismutase, catalase, glutathione reductase, inorganic peroxides and glutathione in the adult housefly. Comp Biochem Physiol 78C:283–288

Anisimov VN, Mylnikov SV, Oparina TI, Khavinson VKh (1997) Effect of melatonin and pineal peptide preparation epithalamin on life span and free radical oxidation in *Drosophila melanogaster*. Mech Ageing Dev 97:81–91

Anisimov VN, Mylnikov SV, Khavinson VKh (1998) Pineal peptide preparation epithalamin increases the life span of fruit flies, mice and rats. Mech Ageing Dev 103:123–132

Bartosz G, Leyko W, Fried R (1979) Superoxide dismutase and lifespan of *Drosophila melanogaster*. Experientia 35:1193

Bieschke ET, Wheeler JC, Tower J (1998) Doxycycline-induced transgene expression during *Drosophila* development and aging. Mol Gen Genet 258:571–579

Bitterman N, Skapa E, Gutterman A (1997) Starvation and dehydration attenuate CNS oxygen toxicity in rats. Brain Res 761:146–150

Bohr VA, Anson RM (1995) DNA damage, mutation and fine structure DNA repair in aging. Mutat Res 338:25–34

Brack C, Bechter-Thüring, E, Labuhn M (1997) N-acetylcysteine slows down ageing and increases the life span of *Drosophila melanogaster*. Cell Mol Life Sci 53:960–966

Breimer LH (1983) Urea-DNA glycosylase in mammalian cells. Biochemistry 22:4192–4197

Campbell JA (1936) Body temperature and oxygen poisoning. J Physiol 89:17P–18P

Chance B, Sies H, Boveris A (1979) Hydroperoxide metabolism in mammalian organs. Physiol Rev 59:527–605

Comfort A, Youhotsky-Gore I, Pathmanathan K (1971) Effect of ethoxyquin on the longevity of C3H mice. Nature 229:254–255

Farmer KJ, Sohal RS (1989) Relationship between superoxide anion radical generation and aging in the housefly, *Musca domestica*. Free Radical Biol Med 7:23–29

Fleming JE, Leon HA, Miquel J (1981) Effects of ethidium bromide on development and aging of *Drosophila*: implications for the free radical theory of aging. Exp Gerontol 16:287–293

Fleming JE, Shibuya RB, Bensch KG (1987) Lifespan, oxygen consumption and hydroxyl radical scavenging capacity of two strains of *Drosophila melanogaster*. Age 10:86–89

Gilbert DL, Gerschman R, Fenn WO (1955) Effects of fasting and X-irradiation on oxygen poisoning in mice. Am J Physiol 181:272–274

Grune T, Reinheckel T, Davies KJA (1997) Degradation of oxidized proteins in mammalian cells. FASEB J 11:526–534

Harman D (1956) Aging: a theory based on free radical and radiation chemistry. J Gerontol 11:298–300

Harman D (1978) Free radical theory of aging: nutritional implications. Age 1:145–152

Harman D (1981) The aging process. Proc Natl Acad Sci USA 78:7124–7128

Hart RW, Setlow RB (1974) Correlation between deoxyribonucleic acid excision-repair and life-span in a number of mammalian species. Proc Natl Acad Sci USA 71:2169–2173

Herbert V (1994) The antioxidant supplement myth. Am J Clin Nutr 60:157–158

Hermes-Lima M, Storey JM, Storey KB (1998) Antioxidant defenses and metabolic depression. The hypothesis of preparation for oxidative stress in land snails. Comp Biochem Physiol B 120:437–448

Hirano T, Yamaguchi R, Asami S, Iwamoto N, Kasai H (1996) 8-hydroxyguanine levels in nuclear DNA and its repair activity in rat organs associated with age. J Gerontol A Biol Sci Med Sci 51:B303–B307

Hollstein MC, Brooks P, Linn S, Ames BN (1984) Hydroxymethyluracil DNA glycosylase in mammalian cells. Proc Natl Acad Sci USA 81:4003–4007

Hwang C, Sinskey AJ, Lodish HF (1992) Oxidized redox state of glutathione in the endoplasmic reticulum. Science 257:1496–1502

Ishikawa T, Sakurai J (1986) In vivo studies on age dependency of DNA repair with age in mouse skin. Cancer Res 46:1344–1348

Izmaylov DM, Obukhova LK (1996) Geroprotector efficiency depends on viability of control population: lifespan investigation in *Drosophila melanogaster*. Mech Ageing Dev 91:155–164

Izmaylov DM, Obukhova LK (1999) Geroprotector effectiveness of melatonin: investigation of lifespan of *Drosophila melanogaster*. Mech Ageing Dev 106:233–240

Jazwinski SM (1996) Longevity, genes, and aging. Science 273:54–59

Johnson TE, Shook D, Murakami S, Cypser J (1999) Increased resistance to stress is a marker for gerontogenes leading to increased health and longevity in nematodes. In: Bohr VA, Clark BFC, Stevnsner T (eds) Molecular biology of aging. Munksgaard, Copenhagen, pp 25–34

Koga T, Takumi K (1995) Nutrient starvation induces cross protection against heat, osmotic, or H_2O_2 challenge in *Vibrio parahaemolyticus*. Microbiol Immunol 39:213–215

Kohn RR (1971) Effect of antioxidants on life span of C57BL mice. J Gerontol 26:378–380

Ku H-H, Brunk UT, Sohal RS (1993) Relationship between mitochondrial superoxide and hydrogen peroxide production and longevity of mammalian species. Free Radical Biol Med 15:621–627

Lamb MJ, McDonald RP (1973) Heat tolerance changes with age in normal and irradiated *Drosophila melanogaster*. Exp Gerontol 8:207–217

Lee YJ, Ducoff HS (1984) Radiation-enhanced resistance to oxygen: a possible relationship to radiation-enhanced longevity. Mech Ageing Dev 27:101–109

Lin Y-J, Seroude L, Benzer S (1998) Extended life-span and stress resistance in the *Drosophila* mutant *methuselah*. Science 282:943–946

Loeb J, Northrop J (1917) On the influence of food and temperature upon the duration of life. J Biol Chem 32:103–121

Manton KG, Stallard E, Wing S (1991) Analyses of black and white differentials in the age trajectory of mortality in two closed cohort studies. Stat Med 10:1043–1059

Massie HR, Williams TR (1979) Increased longevity of *Drosophila melanogaster* with lactic and gluconic acids. Exp Gerontol 14:109–115

Massie HR, Baird MB, Piekielniak MJ (1976) Ascorbic acid and longevity in *Drosophila*. Exp Gerontol 11:37–41

Massie HR, Aiello VR, Williams TR, Baird MB, Hough JL (1993) Effect of vitamin A on longevity. Exp Gerontol 28:601–610

McArthur MC, Sohal RS (1982) Relationship between metabolic rate, aging, lipid peroxidation and fluorescent age pigment in milkweed bug, Oncopeltus fasciatus (hemiptera). J Gerontol 37:268–274

McCord JM (1995) Superoxide radical: controversies, contradictions, and paradoxes. Proc Natl Acad Sci USA 209:112–117

Miquel J, Economos AC (1979) Favorable effects of the antioxidants sodium and magnesium thiazolidine carboxylate on the vitality and life span of *Drosophila* and mice. Exp Gerontol 14:279–285

Miquel J, Lundren PR, Bensch KG, Atlan H (1976) Effects of temperature on the life span, vitality and fine structure of *Drosophila melanogaster*. Mech Ageing Dev 5:347–370

Miquel J, Fleming J, Economos AC (1982) Antioxidants, metabolic rate and aging in *Drosophila*. Arch Gerontol Geriatr 1:159–165

Miquel J, Binnard R, Fleming JE (1983) Role of metabolic rate and DNA-repair in *Drosophila* aging: implications for the mitochondrial mutation theory of aging. Exp Gerontol 18:167–171

Mockett RJ, Sohal RS, Orr WC (1999) Overexpression of glutathione reductase extends survival in transgenic *Drosophila melanogaster* under hyperoxia but not normoxia. FASEB J 13:1733–1742

Newton RK, Ducore JM, Sohal RS (1989) Relationship between life expectancy and endogenous DNA single-strand breakage, strand break induction and DNA repair capacity in the adult housefly, *Musca domestica*. Mech Ageing Dev 49:259–270

Nohl H, Hegner D (1978) Do mitochondria produce oxygen radicals in vivo? Eur J Biochem 82:563–567

Omar BA, Gad NM, Jordan MC, Striplin SP, Russell WJ, Downey JM, McCord JM (1990) Cardioprotection by Cu,Zn-superoxide dismutase is lost at high doses in the reoxygenated heart. Free Radical Biol Med 9:465–471

Orr WC, Sohal RS (1992) The effects of catalase gene overexpression on life span and resistance to oxidative stress in transgenic *Drosophila melanogaster*. Arch Biochem Biophys 297:35–41

Orr WC, Sohal RS (1993) Effects of Cu-Zn superoxide dismutase overexpression on life span and resistance to oxidative stress in transgenic *Drosophila melanogaster*. Arch Biochem Biophys 301:34–40

Orr WC, Sohal RS (1994) Extension of life-span by overexpression of superoxide dismutase and catalase in *Drosophila melanogaster*. Science 263:1128–1130

Orr WC, Arnold LA, Sohal RS (1992) Relationship between catalase activity, life span and some parameters associated with antioxidant defenses in *Drosophila melanogaster*. Mech Ageing Dev 63:287–296

Oudes AJ, Herr CM, Olsen Y, Fleming JE (1998) Age-dependent accumulation of advanced glycation end-products in adult *Drosophila melanogaster*. Mech Ageing Dev 100:221–229

Pacifici RE, Davies KJA (1991) Protein, lipid and DNA repair systems in oxidative stress: the free-radical theory of aging revisited. Gerontology 37:166–180

Parkes TL, Elia AJ, Dickinson D, Hilliker AJ, Phillips JP, Boulianne GL (1998) Extension of *Drosophila* lifespan by overexpression of human *SOD1* in motorneurons. Nat Genet 19: 171–174

Parmar MKB, Machin D (1995) Survival analysis: a practical approach. John Wiley and Sons, Chichester

Parsons PA (1996) Rapid development and a long life: an association expected under a stress theory of aging. Experientia 52:643–646

Pearl R (1928) The rate of living. Alfred A Knopf, New York

Pennisi E (1998) Single gene controls fruit fly life-span. Science 282:856

Ragland SS, Sohal RS (1973) Mating behavior, physical activity, and aging in the housefly, *Musca domestica*. Exp Gerontol 8:135–145

Reveillaud I, Niedzwiecki A, Bensch KG, Fleming JE (1991) Expression of bovine superoxide dismutase in *Drosophila melanogaster* augments resistance to oxidative stress. Mol Cell Biol 11:632–640

Reveillaud I, Phillips J, Duyf B, Hilliker A, Kongpachith A, Fleming JE (1994) Phenotypic rescue by a bovine transgene in a Cu,Zn-superoxide dismutase-null mutant of *Drosophila melanogaster*. Mol Cell Biol 14:1302–1307

Rubner M (1908) Das Problem der Lebensdauer und seine Beziehungen zum Wachstum und Ernährung. Oldenburg, München

Ruddle DL, Yengoyan LS, Miquel J, Marcuson R, Fleming JE (1988) Propyl gallate delays senescence in *Drosophila melanogaster*. Age 11:54–58

Schmidt-Nielsen K (1984) Scaling: why is animal size so important? Cambridge University Press, Cambridge

Seto NOL, Hayashi S, Tener GM (1990) Overexpression of Cu,Zn-superoxide dismutase in *Drosophila* does not affect life span. Proc Natl Acad Sci USA 87:4270–4274

Sohal RS (1982) Oxygen consumption and life span in the adult male housefly, *Musca domestica*. Age 5:21–24

Sohal RS (1986) The rate of living theory: a contemporary interpretation. In: Collatz KG, Sohal RS (eds) Insect aging. Springer, Berlin Heidelberg New York, pp 23–44

Sohal RS (1988) Effect of hydrogen peroxide administration on life span, superoxide dismutase, catalase and glutathione in the adult housefly, *Musca domestica*. Exp Gerontol 23:211–216

Sohal RS, Allen RG (1986) Relationship between oxygen metabolism, aging and development. Adv Free Radical Biol Med 2:117–160

Sohal RS, Brunk UT (1990) Lipofuscin as an indicator of oxidative stress. In: Porta EA (ed) Lipofuscin and ceroid pigments. Plenum, New York, pp 17–29

Sohal RS, Dubey A (1994) Mitochondrial oxidative damage, hydrogen peroxide release, and aging. Free Radical Biol Med 16:621–626

Sohal RS, Sohal BH (1991) Hydrogen peroxide release by mitochondria increases during aging. Mech Ageing Dev 57:187–202

Sohal RS, Weindruch R (1996) Oxidative stress, caloric restriction, and aging. Science 273:59–63

Sohal RS, Wolfe LS (1986) Lipofuscin: characteristics and significance. In: Swaab DF, Fliers E, Mirmiran M, Van Gool WD, Van Haaren F (eds) Progress in brain research, 70. Elsevier, Amsterdam, pp 171–183

Sohal RS, Farmer KJ, Allen RG, Cohen NR (1984a) Effect of age on oxygen consumption, superoxide dismutase, catalase, glutathione, inorganic peroxides and chloroform-soluble antioxidants in the adult male housefly, *Musca domestica*. Mech Ageing Dev 24:185–195

Sohal RS, Farmer KJ, Allen RG, Ragland SS (1984b) Effect of diethyldithiocarbamate on life span, metabolic rate, superoxide dismutase, catalase, inorganic peroxides and glutathione in the adult housefly, *Musca domestica*. Mech Ageing Dev 24:175–183

Sohal RS, Allen RG, Farmer KJ, Newton RK, Toy PL (1985a) Effects of exogenous antioxidants on the levels of endogenous antioxidants, lipid-soluble fluorescent material and life-span in the housefly, *Musca domestica*. Mech Ageing Dev 31:329–336

Sohal RS, Muller A, Koletzko B, Sies H (1985b) Effect of age and ambient temperature on *n*-pentane production in adult housefly, *Musca domestica*. Mech Ageing Dev 29:317–326

Sohal RS, Svensson I, Sohal BH, Brunk UT (1989) Superoxide anion radical production in different animal species. Mech Ageing Dev 49:129–135

Sohal RS, Arnold L, Orr WC (1990a) Effect of age on superoxide dismutase, catalase, glutathione reductase, inorganic peroxides, TBA-reactive material, GSH/GSSG, NADPH/NADP$^+$ and NADH/NAD$^+$ in *Drosophila melanogaster*. Mech Ageing Dev 56:223–235

Sohal RS, Arnold LA, Sohal BH (1990b) Age-related changes in antioxidant enzymes and prooxidant generation in tissues of the rat with special reference to parameters in two insect species. Free Radical Biol Med 10:495–500

Sohal RS, Sohal BH, Brunk UT (1990c) Relationship between antioxidant defenses and longevity in different mammalian species. Mech Ageing Dev 53:217–227

Sohal RS, Svensson I, Brunk UT (1990d) Hydrogen peroxide production by liver mitochondria in different species. Mech Ageing Dev 53:209–215

Sohal RS, Agarwal S, Dubey A, Orr WC (1993) Protein oxidative damage is associated with life expectancy. Proc Natl Acad Sci USA 90:7255–7259

Sohal RS, Agarwal S, Candas M, Forster M, Lal H (1994a) Effect of age and caloric restriction on DNA oxidative damage in different tissues of C57BL/6 mice. Mech Ageing Dev 76:215–224

Sohal RS, Ku H-H, Agarwal S, Forster MJ, Lal H (1994b) Oxidative damage, mitochondrial oxidant generation and antioxidant defenses during aging and in response to food restriction in the mouse. Mech Ageing Dev 74:121–133

Sohal RS, Agarwal A, Agarwal S, Orr WC (1995a) Simultaneous overexpression of copper- and zinc-containing superoxide dismutase and catalase retards age-related oxidative damage and increases metabolic potential in *Drosophila melanogaster*. J Biol Chem 270:15671–15674

Sohal RS, Sohal BH, Orr WC (1995b) Mitochondrial superoxide and hydrogen peroxide generation, protein oxidative damage, and longevity in different species of flies. Free Radical Biol Med 19:499–504

Starke-Reed PE, Oliver CN (1989) Protein oxidation and proteolysis during aging and oxidative stress. Arch Biochem Biophys 275:559–567

Sun J, Tower J (1999) FLP Recombinase-mediated induction of Cu/Zn-superoxide dismutase transgene expression can extend the lifespan of adult *Drosophila melanogaster* flies. Mol Cell Biol 19:216–228

Tatar M, Khazaeli AA, Curtsinger JW (1997) Chaperoning extended life. Nature 390:30

Teixeira HD, Schumacher RI, Meneghini R (1998) Lower intracellular hydrogen peroxide levels in cells overexpressing CuZn-superoxide dismutase. Proc Natl Acad Sci USA 95:7872–7875

The Alpha-Tocopherol, Beta-Carotene Cancer Prevention Study Group (1994) The effect of vitamin E and beta carotene on the incidence of lung cancer and other cancers in male smokers. N Engl J Med 330:1029–1035

Van Voorhies WA, Ward S (1999) Genetic and environmental conditions that increase longevity in *Caenorhabditis elegans* decrease metabolic rate. Proc Natl Acad Sci USA 96:11399–11403

Wheeler JC, Bieschke ET, Tower J (1995) Muscle-specific expression of *Drosophila* hsp70 in response to aging and oxidative stress. Proc Natl Acad Sci USA 92:10408–10412

Yong-Xing M, Yue Z, Chuan-Fu W, Zan-Shun W, Su-Ying C, Mei-Hua S, Jie-Ming G, Jian-Gang Z, Qi G, Lin H (1997) The aging retarding effect of 'Long-Life CiLi'. Mech Ageing Dev 96:171–180

Yu BP (1996) Aging and oxidative stress: modulation by dietary restriction. Free Radical Biol Med 21:651–668

Regulation of Gene Expression During Aging

Stephen L. Helfand and Blanka Rogina

1
Importance of Examining Gene Expression During Aging

The phenomenon of aging is generally believed to be genetically determined and environmentally modulated. Although genetics plays a major role in the determination of life span, the mechanisms by which genetic elements determine life span are poorly understood. Models range from specific highly deterministic genetically based programs to genetically defined constitutions (Finch 1990). One of the major foci in modern molecular gerontology is the mechanistic role that the genome plays in the aging process and life span.

Since the principal mechanism by which the genome exerts its control is through the regulation of gene expression, it is likely that the control of gene regulation plays a central role in the aging process. The subtle stereotypic changes that occur during adult life may therefore be reflected in the pattern of gene expression. The examination of bulk changes in the synthesis of total RNA and protein have been documented in both vertebrate and invertebrate model systems (Smith et al. 1970; Baker et al. 1985; Webster 1985; Levenbook 1986; Remmen et al. 1995). More recently, new molecular genetic techniques have been developed which permit a more detailed assessment of gene expression during aging. Nowhere has this been more extensively studied than in the model system *Drosophila melanogaster*.

2
Drosophila as a Model System for Studying Gene Expression During Aging

The advantages of using *Drosophila* for studying aging are numerous and include: (1) a relatively short life span and ease of maintenance; (2) adult cells

Department of BioStructure and Function, University of Connecticut Health Center, 263 Farmington Avenue, Farmington, CT 06030, USA

Results and Problems in Cell Differentiation, Vol. 29
Hekimi (Ed.): The Molecular Genetics of Aging
© Springer-Verlag Berlin Heidelberg 2000

that are postmitotic and fully differentiated (except for some gonadal and gut cells), eliminating confounding problems from cell replacement (Smith 1962; Bozuck 1972; Mayer and Baker 1985; Ito and Hotta 1992); (3) powerful molecular and genetic techniques; (4) abundant information and accessible manipulation of genetic and environmental factors in aging fruit flies (Baker et al. 1985; Arking and Dudas 1989; Finch 1990); (5) availability of fly lines with specific gene alterations required to facilitate aging studies; (6) simple and well-studied anatomy, allowing for consistent identification of specific individual cells in every animal; and (7) the techniques of enhancer trap and reporter genes which allow for the qualitative and quantitative examination of gene expression in the whole animal (O'Kane and Gehring 1987; Bier et al. 1989; Freeman 1991).

An additional note concerning the use of *Drosophila* as a model system relates to its relevance to humans and other vertebrates. One of the most amazing features that has emerged from the work on development in *Drosophila* is that many of the underlying molecular mechanisms of development in *Drosophila* turn out to be remarkably similar to those used in humans. Not only are many of the master regulatory molecules structurally homologous between these two systems, but in a series of extraordinary experiments, the use of the human homologue in the fly has shown that they appear to be functionally interchangeable (McGinnis and Krumlauf 1992; Halder et al. 1995). Aging, a similarly complex biological phenomenon, is likely to also share some features between humans and flies.

3
Enhancer Trap and Reporter Gene Techniques Can Be Used to Study Gene Expression During Aging

Enhancer-trap and reporter gene constructs have been used to examine gene expression during aging in *Drosophila* (Helfand et al. 1995; Wheeler et al. 1995). In both the enhancer-trap and reporter gene system the expression of a readily visualizable reporter protein mimics the pattern of expression of the native gene, thus allowing for the assessment of the transcriptional activity, both spatially and temporally, of any marked gene. In many cases, the reporter protein expressed is the bacterial β-galactosidase (β-gal) protein.

Several unique characteristics of the adult fly make these reporter gene marked genes useful for examination of both the temporal and spatial pattern of gene expression during aging. For most tissues, the expression of the reporter protein, β-gal, is an accurate indicator of changes in gene activity. Because somatic cells are not added during adult life, except in the gonad and gut (Bozuck 1972; Ito and Hotta 1992) and are not lost in many tissues, such as the antenna (Helfand and Naprta 1996) and eye (Leonard et al. 1992), changes in expression of β-gal over time reflect an alteration in transcriptional activity of mature aging cells and are not caused by gain or loss in cell populations. Also studies on protein synthesis using β-gal have shown that the cellular machinery needed for protein synthesis remains intact throughout the aging process in *Drosophila melanogaster* (Helfand and Naprta 1996). So, decreases in the amount of β-gal expressed during aging from a particular reporter marked gene are not due to the loss of the ability of cells to synthesize β-gal. Furthermore, the β-gal protein degrades rapidly enough for it to accurately indicate expression during aging. Estimates of the rate of degradation of β-gal in the adult fly are on the order of hours in both young and old *Drosophila melanogaster* and thus rapid enough to be used to measure changes in transcriptional activity taking place over days (Helfand and Naprta 1996). Taken together, the maintenance of a stable cell population (no somatic cell division and few cells lost with age in the antenna and eye); preservation of protein synthesis with age; and rapid rate of degradation of β-gal indicate that this system can be used to measure changes in the transcriptional activity of single genes during adult life.

Furthermore, the use of reporter proteins, such as β-galactosidase and Green Fluorescent Protein, in *Drosophila* allows for the examination of gene expression in the intact animal. Thus, through the use of these reporter proteins it is possible to examine the temporal and spatial pattern of expression of single genes at the level of individual or small subsets of cells in the adult, a precision not previously available. These techniques provide a means of assessing transcriptional activity during aging which is distinct from other methods that may reflect changes at other points in the biosynthetic pathway.

4
The Level of Expression of Many Genes Is Dynamically Changing During Adult Life in *Drosophila melanogaster*

Measuring the transcriptional activity of a number of different genes throughout the life of the adult fly, by using the enhancer-trap technique, has revealed the dynamic nature of gene expression. In one study the level of β-gal expression was examined for 49 different enhancer-trap marked genes (Helfand et al. 1995). In 80% of these genes the level of gene expression changed dramatically with age. Many, 55%, of these marked genes showed an increase in the level of transcriptional activity with age. Figure 1 shows the data from four different enhancer-trap lines illustrating the dynamic nature of gene expression throughout the life span of the adult fly.

Line
302

1059

1020

2216

Days   1   5   10   20·25 35·40 50   60

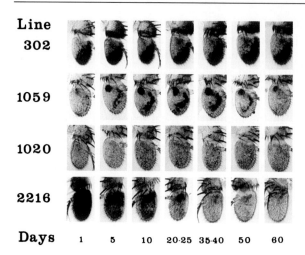

Fig. 1. Expression of a reporter gene shows the dynamic nature of gene expression during adult life. Photomicrographs of whole mount adult antennae from enhancer trap lines *302*, *1059*, *1020*, and *2216* at different ages reacted with X-gal to reveal blue staining in the nuclei of cells that are expressing β-gal. Each photograph represents a typical example from over 60 different antennae examined for each time point. Days listed on the bottom are from the time of the adults' emergence from the pupal case (Helfand et al. 1995)

5
Gene Expression Is Carefully Regulated During Adult Life in *Drosophila melanogaster*

A quantitative method for determining the intensity of X-gal reaction (a measure of β-gal activity) was developed in order to more fully evaluate the temporal pattern of gene expression for the enhancer-trap and reporter gene lines. This is an optically based computer-assisted video system which is able to measure the intensity of X-gal reaction present in each antenna, thus providing a relative measure of β-gal expression (Blake et al. 1995; Helfand et al. 1995). Figure 2 illustrates the use of this system for determining the time-dependent nature of gene expression and shows the diversity of temporal patterns seen.

In addition to showing that gene expression in the adult fly is more dynamic than previously thought, these studies also demonstrate that gene regulation continues to be active throughout the life of the adult fly. Although the expression of different genes may show different temporal patterns, under similar conditions each gene's temporal pattern of expression is stereotypic. These data illustrate that time-dependent mechanisms regulate the expression of many different genes in the adult fly.

6
Some Genes Are Regulated by Mechanisms That Are Linked to Life Span and May Serve as Biomarkers of Aging

Examination of the temporal pattern of expression of many different genes demonstrates the dynamic and well-regulated nature of gene expression in the

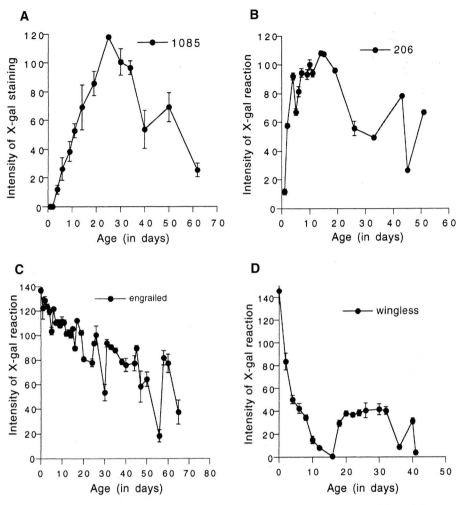

Fig. 2A–D. Quantitative analysis of β-gal staining throughout the life span of four different marked genes shows the diversity and well-regulated nature of gene expression. The relative staining of β-gal was determined by using a computer-assisted video system on X-gal-stained antennae. Antennae from the **A** 1085 and **B** 206 enhancer trap lines. Antennae from enhancer trap inserts marking the **C** *engrailed* and **D** *wingless* genes. Each *point* represents the average and Standard Error of the Mean (SEM) of ten different male antennae (Helfand et al. 1995; Rogina and Helfand 1997)

adult fly. What is the relationship between these time-dependent patterns of expression and life span? One way to try to understand this relationship is to examine gene expression after altering life span. Environmental and genetic methods can be used to change life span in *Drosophila melanogaster*. As poikilotherms, the life span of these animals can be changed by altering the ambient temperature at which they are living. A change in temperature of 11°C results in a nearly three-fold difference in life span. For a typical wild-type strain such as Canton-S, maximal life span at 29°C is ~45 days, at 25°C it is ~75 days, and at 18°C it is ~135 days (see Fig. 3A). Life span can also be altered by genetic mutations such as the X-linked *Shaker*[5] and *Hyperkinetic*[1] which are thought to shorten life span by accelerating the rate of aging (Trout and Kaplan 1970).

When the temporal pattern of expression of some genes is examined in animals living at different ambient temperatures or crossed into the back-

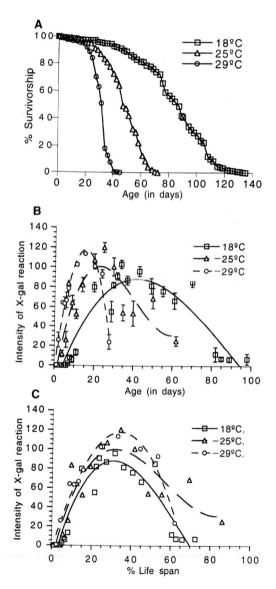

Fig. 3A–C. The temporal pattern of gene expression in the 1085 line scales to life span when life span is altered by ambient temperature. The figures *above* show the effect of altering life span by means of ambient temperature on the temporal pattern of expression of β-gal in the antenna of adult flies from line 1085 at 29°C (○), 25°C (△), and 18°C (□). **A** Life span of the 1085 line at 29, 25, or 18°C. **B** Expression of β-gal throughout the life span of the 1085 line living at 29, 25, or 18°C. The timing of gene expression changes dramatically in calendar time when the animals are cultured at these different ambient temperatures. **C** When viewed with respect to percent life span the timing of gene expression appears to be the same. Animals were cultured throughout development at 25°C and then transferred to either 29, 25, or 18°C upon emergence as adults. A computer-assisted video system was used to determine the intensity of the X-gal reaction in each specimen. At least 20 different antennae were sampled for each point. The *error bars* are standard errors of the mean (SEM). The *curves* were fit to the points for each temperature using a third-degree polynomial equation. (Helfand et al. 1995)

ground of *Shaker*[5] or *Hyperkinetic*[1] mutations, significant changes in the temporal patterns are seen. Examination of these data demonstrates that the overall pattern of gene expression has not changed, but the timing has been altered. This can be illustrated by examining the effects of ambient temperature on the expression of the 1085 gene. Figure 3 shows the results of growing animals at three different ambient temperatures on the temporal pattern of expression of the 1085 marked gene. Similar changes are seen when life span is altered using the *Shaker*[5] or *Hyperkinetic*[1] mutations (Helfand et al. 1995; Rogina and Helfand 1995).

When animals are grown at 29, 25, or 18°C the temporal pattern of expression of the 1085 gene is dramatically altered (Fig. 3B). Expression at 29°C occurs earliest, by day 2, reaches its peak first, by day 15, and has a precipitous decline. At the other extreme, 1085 expression at 18°C is not initiated until day 7, does not peak until after day 35, and shows a slow decline afterwards. Normalizing 1085 gene expression to the different life spans reveals an important clue to its possible regulation. When gene expression is plotted as a function of percent life span rather than calendar time, the three different temporal patterns superimpose (Fig. 3C). The temporal patterns of expression change in a manner that is proportional to the alteration in life span. It appears that for genes such as 1085, changes in the temporal pattern of gene expression are compensating for changes in the rate of aging or life span. Studies with the *Shaker*[5] or *Hyperkinetic*[1] mutations also show this same scaling with respect to life span (Rogina and Helfand 1995). This shows that the regulation of expression of the 1085 gene during adult life scales to life span, implying that there is some linkage between the regulation of expression of the 1085 gene and life span. Furthermore, these genes, whose pattern of expression are more closely linked to life span than calendar time, may serve as biomarkers of aging.

7
The Expression of Some Genes Is Not Changed by Environmental or Genetic Manipulations That Alter Life Span

Not all genes appear to be regulated by mechanisms linked to life span. The temporal pattern of expression for these genes does not appear to be linked to physiological age (Rogina and Helfand 1996). Several genes have been identified which appear to define a class of genes whose temporal pattern of expression is independent of the rate of aging, as altered by ambient temperature or mutations in *Shaker* or *Hyperkinetic*. One such gene, called 2216, shows a temporally regulated pattern of gene expression in the adult defined by a nearly steady decline from a high level of expression upon emergence of the adult from the pupal case until near undetectable levels by 40 days of life at 25°C. Remarkably, the rate of decline in the level of expression of 2216 is unresponsive to ambient temperature or to genetic manipulations that are thought to lead to an accelerated rate of aging (e.g., *Hk*, *Sh*: Rogina and Helfand 1996; Trout and Kaplan 1970).

As shown in Fig. 4A the rate of decline in amount of β-gal expression with the 2216 reporter gene marked gene is indistinguishable at 18, 25, or 29°C, despite a threefold difference in expected life span. This is further illustrated by the accompanying figure (Fig. 4B) which plots the level of expression of β-gal against percent life span for each of these three conditions. Under circumstances in which the level of expression is linked to life span the three lines would converge as is seen with line 1085 (see Fig. 3). With 2216 they are found to diverge, indicating that the temporal pattern of expression is not following physiological age. Instead, the 2216 gene appears to be responding to some mechanism that keeps track of chronological or calendar time (Rogina and Helfand 1996).

8
Use of Temporal Patterns of Gene Expression as Biomarkers of Aging

Genes whose temporal patterns of expression scale to life span may be useful as biomarkers of aging in the analysis of manipulations which alter life span. These biomarkers of aging can be used to assess whether a particular

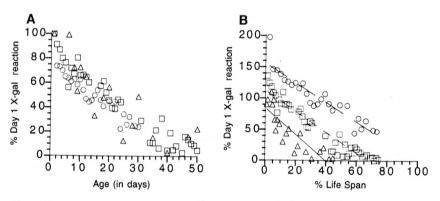

Fig. 4A,B. A gene whose temporal pattern of expression appears independent of ambient temperature. **A** Intensity of X-gal reaction is plotted as a percent of that at day 1 for each of the three temperatures 29°C (○), 25°C (□) and 18°C (△). **B** Intensity of the X-gal reaction plotted against percent life span. The percent life span for each value was determined by dividing the day for each point by the maximum life span of line 2216 at that temperature. The *curves* were fit to the points for each temperature 29°C (○), 25°C (□) and 18°C (△) using a linear equation (Rogina and Helfand 1996)

intervention, environmental or genetic, that shortens or lengthens life span, is associated with a change in the rate of aging. An illustration of this is the use of gene expression as biomarkers in the analysis of the short life span of the X-linked *Drosophila* mutation, *drop-dead* (Rogina et al. 1997).

Animals mutant for *drop-dead* all die by day 15 of adult life. The etiology of the shortened adult life span of the *drop-dead* animals is not known. One possible explanation is that the product of the *drop-dead* gene provides an essential element for the development and/or maintenance of the young adult. A lack of *drop-dead* could thus lead to a catastrophic event and the early death of the adult animal. If this is the case, then it would be expected that the short life span of the *drop-dead* animals would be associated with a truncation in the normal temporal pattern of our molecular biomarkers. Alternatively, if the *drop-dead* mutation somehow resulted in an acceleration of the aging process, then an accelerated sequence of gene expression should be seen with the biomarkers. Examination of three different biomarkers showed just such an acceleration in their temporal pattern of expression (Rogina et al. 1997). An example of the accelerated expression pattern of one of these biomarkers, the *wingless* gene, is shown in Fig. 5. These studies suggest that the early demise of the *drop-dead* animal may be related to a rapid increase in the process of aging.

9
The *drop-dead* Mutation May Be Used to Accelerate Screens for Long-Lived Mutations

Although it is not yet understood how mutations in the *drop-dead* gene set into motion a series of events resulting in an acceleration of the aging process, knowledge that *drop-dead* animals show characteristics of rapid aging makes them valuable for use in future genetic studies of aging. A major problem that has delayed the isolation of long-lived mutants in *Drosophila* is procedural, stemming from the same technical constraint: no accepted surrogate phenotype for predicting longevity, other than survivorship, is available. Screening for possible long-lived mutants using survivorship takes a long time. Under standard culturing conditions, the maximal life span of Canton-S is typically ~80 days. The other common *Drosophila* strain, Oregon-R, lives even longer, usually up to ~90–100 days. To identify a mutant line that is long-lived would require survivorship tests taking over 3 to 4 months for each potential mutant line. Although life span can be shortened to 45–50 days at higher temperatures such as 29°C, temperatures such as these are not physiologic for the animals and introduce other confounding problems.

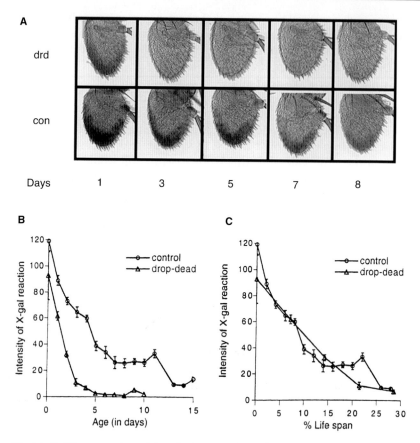

Fig. 5A–C. The expression of the *wingless* gene is accelerated in proportion to the shortened life span of *drop-dead* animals. **A** Photomicrographs of whole-mount adult antennae from animals containing one autosomal copy of the *wg* enhancer-trap chromosome, with or without *drop-dead* on the X-chromosome, reacted with X-gal to reveal dark staining in cells expressing β-gal. Each is a typical example from over ten flies at each age after emergence of the adult. **B** Expression of *wingless* decreases gradually during aging in control (○) antennae and rapidly in *drop-dead* antennae (△). **C** Expression of *wingless* as a percentage of life span. Symbols as in B. The *error bars* are standard errors of the mean (SEM) (Rogina et al. 1997)

The use of the rapidly aging *drop-dead* mutant greatly reduces survivorship and thus the time it takes to screen for long-lived mutants. In a *drop-dead* background mean life span is ~4 days and maximum ~15 days, as opposed to ~35–40 days and ~80 days in a Canton-S wild-type background (Fig. 6). In the background of *drop-dead*, any mutagenized animals that lived to day 20 could be carrying a mutation leading to an extension in life span. This would provide a rapid enrichment for possible long-lived mutants.

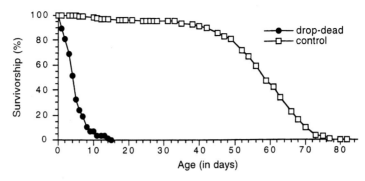

Fig. 6. Survivorship for *drop-dead* males (●) and matched controls (□) at 25°C. (Rogina et al. 1997)

10
Studies on Gene Expression Suggest
that Not All Things Fall Apart During Aging

It has generally been thought that with age all things fall apart. One of the implications of these ideas is that with aging, homeostasis will be lost globally, and a general increase in dysregulation will occur. Examination of age-specific variability in gene expression, a sensitive physiological measure, fails to support this idea (Fig. 7). The variation in gene expression of six different genes was studied over the life span of the adult fly. These studies showed that there is no systematic increase (or decrease) in the variability of gene expression with age (Rogina et al. 1998). Despite attempts to exacerbate variability by culturing animals at suboptimal temperatures such as 29°C or in a variety of different genetic backgrounds, no consistent increase in variability of gene expression was noted as a consequence of age. Some of the genes examined did have a great deal of variability among individuals of the same age, but this did not increase as a function of age. By examining gene expression it was shown that, although regulation of gene expression is an important element of life, a global loss in the control of gene expression appears to be neither a cause nor a consequence of the process of aging.

11
Conclusions

The examination of gene expression during aging is leading to changes in the way in which the process of aging is being viewed. In contrast to the idea that the period of adult life is one of stasis or passive decline, studies of gene

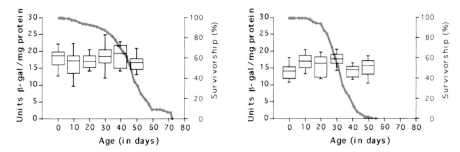

Fig. 7. The *boxes* on this graph represent the variations in expression of the fly opsin 1 photoreceptor gene seen at a particular age from a cohort of ten male (*left*) and ten female (*right*) adult flies. Expression of the opsin 1 photoreceptor gene was determined by measuring the level of the β-gal reporter from an Rh1 opsin reporter gene construct (Mismer and Rubin 1987). The *line in the background* is a survivorship curve for each population. If all things fall apart together with age, then it would be expected that an increase in variation would be seen. This figure does not show a tendency towards an increase in variation with age. A decrease in the efficiency of gene regulation as a function of age is not seen (Rogina et al. 1998)

expression show that adult life is characterized by the same dynamic highly regulated changes that occur during earlier periods of life. The dynamic changes in gene expression seen demonstrate that gene regulation remains carefully controlled throughout all adult life. The changes in gene expression do not appear to be random. Examination of the mechanisms underlying the regulation of gene expression during adult life shows that the regulation of some genes is through mechanisms that are somehow linked to life span. The regulation of these genes thus appears to be more closely linked to the physiological age of the animal than to its chronological age and may serve as biomarkers of aging or biopredictors of longevity. Other genes have also been identified in which regulation of their temporal pattern of expression appears to be through mechanisms that are independent of life span or physiological age. These genes appear to be regulated by mechanisms more closely linked to chronological age or calendar time. The finding of stereotypic changes in gene expression linked to life span or calendar age suggests the possibility that important timing mechanisms are involved in regulating gene expression during adult life. Studies of gene expression have also begun to show the inadequacy of some of our common sense notions about a global loss of homeostasis during aging. Although the aging process may be characterized as a state of deterioration extending over most of adult life, gene expression, a fundamental element of life, strikingly appears to continue to be well regulated. Aging may be associated with the loss of homeostasis of some physiological elements but a loss of regulation of gene expression is not one of these. The examination of the control of gene expression during aging provides new entrances into our understanding of the role of genetic elements in aging. The finding of carefully regulated dynamic changes in gene expression during aging has begun to redirect and reinvigorate our search for a molecular genetic understanding of the process of aging.

Acknowledgments. This work was supported by grants from the National Institute on Aging (AG14532 and AG16667), National Institutes of Health-supported (NIA) Claude Pepper Older Americans Independence Center at the University of Connecticut Center on Aging, and the American Federation for Aging Research.

References

Arking R, Dudas SP (1989) Review of genetic investigations into the aging processes of *Drosophila*. JAGS 37:757–773

Baker GT, Jacobson M, Mokrynski G (1985) "Aging in *Drosophila*". In: Adelman RC, Roth GS (eds) Aging in Drosophila. CRC Press, Boca Raton, FL

Bier E, Vassin H, Shepherd S, Lee K, McCall K, Barbel S, Ackerman L, Carretto L, Uemura T, Greel E, Jan L, Jan Y (1989) Searching for pattern and mutation in the *Drosophila* genome with a P-*lacZ* vector. Genes and Dev 3:1273–1287

Blake K, Rogina B, Centurion A, Helfand SL (1995) Changes in gene expression during post-eclosional development in the olfactory system of *Drosophila melanogaster*. Mech Dev 52:179–185

Bozuck AN (1972) DNA synthesis in the absence of somatic cell division associated with ageing in *Drosophila subobscura*. Exp Gerontol 7:147–156

Finch CE (1990) Longevity, senescence, and the genome. In: Longevity, senescence, and the genome. University of Chicago Press, Chicago

Freeman M (1991) First, trap your enhancer. Curr Biol 1:378–381

Halder G, Callaerts P, Gehring WJ (1995) Induction of ectopic eyes by targeted expression of the *eyeless* gene in *Drosophila*. Science 267:1788–1792

Helfand SL, Naprta B (1996) The expression of a reporter protein, β-galactosidase, is preserved during maturation and aging in some cells of the adult *Drosophila melanogaster*. Mech Dev 55:45–51

Helfand SL, Blake KJ, Rogina B, Stracks MD, Centurion A, Naprta B (1995) Temporal patterns of gene expression in the antenna of the adult *Drosophila melanogaster*. Genetics 140:549–555

Ito K, Hotta Y (1992) Proliferation pattern of postembryonic neuroblasts in the brain of *Drosophila melanogaster*. Dev Biol 149:134–148

Leonard DS, Bowman VD, Ready DF, Pak WL (1992) Degeneration of photoreceptors in rhodopsin mutants of *Drosophila*. J Neurobiol 23:605–626

Levenbook L (1986) Protein synthesis in relation to insect aging: An overview. Springer, Berlin Heidelberg New York

Mayer PJ, Baker GT (1985) Genetic aspects of *Drosophila* as a model system of eukaryotic aging. Int Rev Cytol 95:61–102

McGinnis W, Krumlauf R (1992) Homeobox genes and axial patterning. Cell 68:283–302

Mismer D, Rubin GM (1987) Analysis of the promoter of the *ninaE* opsin gene in *Drosophila melanogaster*. Genetics 116:565–578

O'Kane K, Gehring W (1987) Detection in situ of genomic regulatory elements in *Drosophila*. Proc Natl Acad Sci USA 84:9123–9127

Remmen HV, Ward WF, Sabia RV, Richardson A (1995) Gene expression and protein degradation. Oxford University Press, London

Rogina B, Helfand SL (1995) Regulation of gene expression is linked to life span in adult *Drosophila*. Genetics 141:1043–1048

Rogina B, Helfand SL (1996) Timing of expression of a gene in the adult *Drosophila* is regulated by mechanisms independent of temperature and metabolic rate. Genetics 143:1643–1651

Rogina B, Benzer S, Helfand SL (1997) *Drosophila drop-dead* mutations accelerate the time course of age-related markers. Proc Natl Acad Sci USA 94:6303–6306

Rogina B, Vaupel JW, Partridge L, Helfand SL (1998) Regulation of gene expression is preserved in aging *Drosophila melanogaster*. Curr Biol 8:475–478

Smith JM (1962) The causes of ageing. Proc R Soc Lond Ser B 157:115–127

Smith JM, Bozuck AN, Tebbutt S (1970) Protein turnover in adult *Drosophila*. J Insect Physiol 16:601–613

Trout WE, Kaplan WD (1970) A relation between longevity, metabolic rate, and activity in Shaker mutants of *Drosophila melanogaster*. Exp Gerontol 5:83–92

Webster GC (1985) Protein synthesis in aging organisms. Raven Press, New York

Wheeler JC, Bieschke ET, Tower J (1995) Muscle-specific expression of *Drosophila* hsp70 in response to aging and oxidative stress. Proc Natl Acad Sci USA 92:10408–10412

Crossroads of Aging in the Nematode *Caenorhabditis elegans*

Siegfried Hekimi

1
Introduction

1.1
Life Span Versus Aging

Aging can be defined in three ways: (1) as a progressive increase in the probability of dying of nonaccidental causes, (2) as a progressive increase in the probability of being afflicted with a number of specific diseases, such as cancer, cardiovascular diseases, and neurodegenerative diseases, and (3) as a progressive increase in the prevalence of features that are not in themselves pathological, but which are linked to chronological age, like wrinkled skin or white hair. In recent years, several investigators have used definition (1) and the measure of life span in the nematode *Caenorhabditis elegans* to study genetic, cellular, and molecular mechanisms that might be responsible for the aging process in all organisms (Hekimi et al. 1998).

1.2
The Worm

There are several clear advantages in using *C. elegans* as a model system to study the genetics of life span in a multicellular organism. (1) Worms are small (the adult is only 1 mm long), and their life cycle, from fertilization to the onset of reproduction, lasts only 3 days (Fig. 1). They can therefore be cultivated in large numbers with a minimum of space and expense. (2) Worms can be frozen indefinitely in liquid nitrogen and revived subsequently. This greatly facilitates genetic screens, as interesting candidate strains do not have to be actively maintained but can be frozen away for later analysis. As strains are not constantly maintained, there is also less risk of unwanted genetic drift and, should such drift occur, it is always possible to

Department of Biology, 1205 Docteur Penfield Avenue, Montréal, PQ, Canada H3A 1B1

Results and Problems in Cell Differentiation, Vol. 29
Hekimi (Ed.): The Molecular Genetics of Aging
© Springer-Verlag Berlin Heidelberg 2000

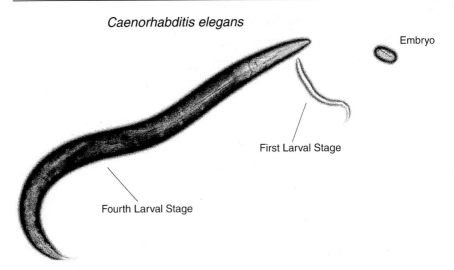

Caenorhabditis elegans

Fig. 1. Three developmental stages of the nematode *Caenorhabditis elegans*. During post-embryonic development there are four larval stages. The fourth larval stage (L4) animals molt into the adult stage and start laying eggs. The L4 animal shown is approximately 800 μm long and as adults can reach up to 1.2 mm long. The first larval stage animal shown is freshly hatched from an egg. Most of the increase in complexity during worm development occurs during embryonic development. Indeed, all cell types and most of the cells present in the adult are already present in the hatching larva

go back to the original frozen isolate. (3) From the beginning of the use of *C. elegans* as a model system, a unique reference strain (N2) has been used. The vast majority of mutations known in worms, including those that affect life span, have been generated in this background thus keeping genetic background effects to a minimum. (4) Worms are normally short-lived: the wild-type reference strain (N2) lives for only 16 days at 20°C (Lakowski and Hekimi 1996). This allows experiments to be concluded within weeks of their initiation. (5) *C. elegans* is an internally self-fertilizing hermaphrodite. This confers several advantages: fully isogenic lines can be studied without the problems associated with inbreeding depression, as the normal mode of existence of this species is to be homozygous at most loci, most of the time. Also, under laboratory conditions, worms do not need to move, either to feed (they live on a lawn of bacteria on which they feed), or to mate (they fertilize their oocytes internally with their own sperm), so that even very debilitated animals can be propagated and studied.

Numerous genetic and molecular tools (not least among them the entire genome sequence) have been developed to clone genes defined only by mutant phenotypes. As will be described below, these tools have now also been applied to the study of aging. As a result, a truly molecular view of the mechanisms involved in life-span determination in this organism is beginning to emerge. Furthermore, the detail in which various aspects of development, growth, and fertility can be quantified in the worm (see Wong et al. 1995 for example), as well as the availability of many mutations that alter these life history parameters (Hekimi et al. 1995, 1998; Lakowski and Hekimi 1996), gives us the opportunity to relate molecular insights to relevant organismal features.

One criticism regarding the use of model systems such as the worm for the study of aging is that one cannot study the pathology of aging. Mostly, we know very little about the physiology of aging worms or why they die. However, as Herndon and Driscoll describe in a chapter of this Volume, new insights into the relationship between life span and various types of degenerative cell death are now being gained through genetic studies in *C. elegans*.

1.3
Three Paths of Longevity

Three genetically and molecularly distinct biological processes that can be altered to lengthen the life span of the worm have been identified: dormancy, the rate of living, and food intake. Below, these processes are first described in detail and the genetic findings that have established their involvement in the aging process. Next, what we know about how these processes interact with each other will be described. Finally, the attempt is made to give a synthetic view of how the three processes participate in shaping those aspects of the biology of the organism that determine its life span.

2
Dormancy

2.1
The Dauer Larva

After embryonic development and hatching, worms proceed through a series a four very similar larval stages (L1–L4) before reaching adulthood. The four types of larvae differ mostly by their size and the extent of development of the somatic gonad and germline. However, under particular environmental conditions, worms will enter the dauer larval stage (from the German *dauern* meaning to last) instead of the L3 larval stage (Fig. 2). The dauer larva is a developmentally arrested dispersal stage that allows worms to survive adverse conditions (Golden and Riddle 1984; Riddle 1988; Riddle and Albert 1997). The mouth and anus of dauer larvae are closed and they are resistant to desiccation. They do not feed and do not move much except when attracted by food or water. When appropriate environmental conditions arise, dauer larvae resume development and proceed to the L4 larval stage and subsequently to adulthood.

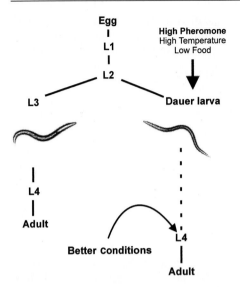

Fig. 2. Two alternative paths of worm development. After embryogenesis, worms can either pass through four similar larval stages and then become adults, or, after the second larval stage, they can become dauer larvae, stop feeding, and arrest development. Dauer larval development is induced by high concentrations of a pheromone that is produced constitutively by worms at all stages and is therefore a measure of the local density of worms. Dauer formation is further promoted by high temperatures and low amounts of food. Dauer larvae do not move unless attracted by food or water. They are more stress resistant than worms at other developmental stages and can live substantially longer than adult worms (up to 6 months for some individuals with a mean around 2 months). Dauer larvae resume development and molt into a fourth stage larva only when they experience better environmental conditions, in particular the presence of food. Therefore, unlike for the other larval stages, the length of the dauer stage is variable (indicated in the figure by a *stippled line*)

Three environmental conditions promote entry into the dauer stage: high temperatures, low amounts of food and, most importantly, high concentration of a pheromone which is constitutively produced by worms at all stages. Development resumes when the pheromone concentration is low and food is present. Entering the dauer larval stage is a decision made for the longterm, as the immediate production of offspring is forfeited. The worm will enter the dauer stage so as to avoid the risk of never completing development or reproducing because of the absence of food. It is logical therefore that the concentration of pheromone is the most important determinant of this decision, as it signals a long-term trend in the environment, that is, the presence of many competitors for resources. High temperature and absence of food (in other words, further adverse conditions) reinforce the response to pheromone.

In the dauer state, worms can survive up to 6 months, which is much longer than the 2–3 weeks that worms live maximally as adults (Riddle 1988; Taub et al. 1999). As dauer larvae have to outlast adverse conditions, it is reasonable to suppose that they are resistant to environmental stresses, and that this property is what allows them to live a long time. Their tissues might sustain less damage from environmental insults or repair damage better. Dauer larvae are probably resistant to internally generated oxidative stress as well, as they have elevated activities of superoxide dismutase (Anderson 1982; Larsen 1993) and catalase (Taub et al. 1999). Furthermore, the expression of *ctl-1*, a gene encoding a cytosolic catalase, is much higher in dauer larvae than in other developmental stages, and a mutation in this gene shortens the life span of dauers (Taub et al. 1999). The issue of the stress resistance of dauer larvae and the effect of stress resistance in general is addressed in more detail in the chapter by Lithgow.

In addition to the long life of animals in the dauer stage, it was observed in a seminal study by Klass and Hirsh (1976) that the life span of worms that have resumed development after passing through the dauer stage is irrespective of the amount of time spent in that stage. Whether worms have spent 2 days or 2 months in the dauer stage, their adult life span and fertility are unaffected. The likely high stress resistance of dauer larvae is not sufficient to explain this observation. Indeed, dauer larvae are not immortal and their probability of dying increases with chronological time (Klass and Hirsh 1976; Kagan et al. 1997; Taub et al. 1999), suggesting that, although more resistant than adults, dauer larvae nonetheless sustain and accumulate damage until the amount of damage becomes lethal. This implies that after a passage through the dauer stage the worms that resume development must deal with the unrepaired damage undergone during dauer life. For example, as worms do not undergo cell replacement and as most somatic cells found in the adult are already present in dauer larvae, any somatic mutation sustained during dauer life will plague the mutant cell throughout the animal's life. It will therefore be interesting to investigate whether post-dauer L4 larvae and adults have better mechanisms of proofreading or higher protein turnover in order to avoid toxic effects from mutant RNA or proteins produced by damaged genes.

2.2
The Genetics of Dauer Formation

Numerous *daf* (*da*uer *f*ormation) genes are active in a complex genetic pathway that is involved in regulating dauer formation (Riddle and Albert 1997; Thomas et al. 1993). These genes are involved in the production of pheromone, the sensing of pheromone, and the transduction of the signal from the sensory cells. Most importantly, some of the genes are likely to be required in the cells that have to choose between dauer and L3 morphology and physiology. There are two types of *daf* mutations, dauer-defective (daf-d) mutations and dauer-constitutive (daf-c) mutations. Two daf-c genes (*daf-2* and *age-1*) and one daf-d mutation (*daf-16*) are most relevant to the discussion of the effect of dormancy on aging (Fig. 3).

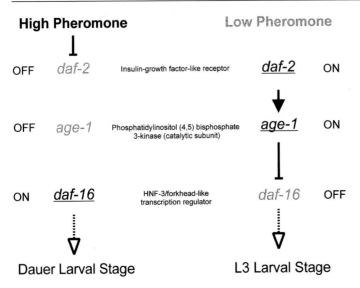

Fig. 3. The genetic pathway determining the two alternative developmental fates in *Caenorhabditis elegans*. High amounts of a pheromone that is produced constitutively by worms at all stages promote the dauer fate. Mutational inactivation of *daf-2* or *age-1* mimics this effect and leads to constitutive dauer formation at high temperature, suggesting that these genes are normally inhibited by pheromone. *daf-16* is necessary for the dauer fate, as mutational inactivation of *daf-16* prevents dauer formation even in the presence of pheromone or mutational inactivation of *daf-2* or *age-1*. All phenotypes of *daf-2* and *age-1* mutations are completely suppressed by mutations in *daf-16*, including their increased life span. *daf-2* encodes an insulin-growth factor-like receptor, *age-1* a phosphatidylinositol (4,5) bisphosphate 3-kinase subunit, and *daf-16* a forkhead-like transcription regulator

   Both *daf-2* and *age-1* were originally defined by temperature-sensitive alleles (Kenyon et al. 1993; Gottlieb and Ruvkun 1994; Malone et al. 1996; Morris et al. 1996; Kimura et al. 1997; Gems et al. 1998). When worms are grown at the nonpermissive temperature (25°C for *daf-2* and 27°C for *age-1*), they become dauer larvae instead of L3 larvae even under conditions that do not normally induce dauer formation (low pheromone, high food). When such dauer larvae are returned to a permissive temperature (e.g., 15°C), most of them will resume development and become fertile adults. On the other hand, when these mutants are grown at the permissive temperature throughout development or when they are switched to the nonpermissive temperature only after the L3 stage, they develop normally, becoming fertile adults without passing through a dauer stage. However, such mutant adults are recognizably different from their wild-type counterparts. For example, their intestines are darker and their movements are sluggish. Their most remarkable feature is that they live much longer than normal adults (Fig. 4). A reasonable interpretation of this observation is that these mutant animals develop some of the characteristics of dauer larvae, including more lipid

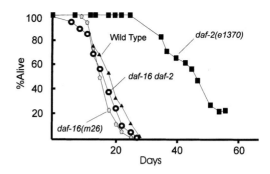

Fig. 4. Death curves for the wild-type (*N2*), *daf-2*, *daf-16*, and *daf-16 daf-2* mutants. *daf-2* mutants live substantially longer than wild-type animals (the curve does not reach 0% alive because the experiment was not terminated at the time of publication). Animals carrying the *daf-16(m26)* mutation have a life span only slightly shorter than the wild-type. However, the *daf-16* mutation entirely suppresses the increase in life span produced by *daf-2* mutation as the *daf-16 daf-2* double mutant animals do not live longer than the wild-type. (After data from Kenyon et al. 1993)

reserves, less motility, and the as yet undefined physiological characteristic (presumably stress resistance) that allows dauer larvae to live a long time (Kenyon et al. 1993).

daf-16, a dauer-defective gene that functions downstream of *daf-2* and *age-1*, suppresses all phenotypes produced by these two genes, including long life (Fig. 4) (Kenyon et al. 1993; Gottlieb and Ruvkun 1994). This means that *daf-16daf-2* and *daf-16age-1* double mutants are dauer-defective even at restrictive temperatures and do not live longer than the wild-type.

2.3
The Molecular Identities of the Dauer Genes

age-1 has been cloned and found to encode a nematode homologue of the p110 subunit of phosphatidylinositol 3-kinase (PI 3-kinase) (Morris et al. 1996). p110 is the catalytic subunit that turns the lipid phosphatidylinositol (4,5)-bisphosphate into phosphatidylinositol (3,4,5)-trisphosphate. PI 3-kinase is part of a very well-studied signal transduction pathway and acts downstream of dimeric growth-factor-receptor tyrosine kinases. Among the types of receptors that have been found upstream of PI 3-kinase are those of the insulin receptor family. Indeed, *daf-2* encodes a nematode member of that family (Kimura et al. 1997). Together, these molecular findings very strongly suggest that *age-1* and *daf-2* participate in a signal transduction cascade similar to that described in vertebrates. Recently, additional genes known to be involved in these pathways such as the PTEN and Akt/PKB have been studied in worms and their homologues there were shown to play a role in dauer formation (Ogg and Ruvkun 1998; Paradis and Ruvkun 1998).

daf-16 has been cloned by two groups (Lin et al. 1997; Ogg et al. 1997) and potentially encodes three isoforms belonging to the hepatocyte nuclear factor 3 (HNF-3)/forkhead family of transcriptional regulators. The finding that *daf-16* encodes a transcription factor suggests that DAF-16 might indeed be the gene product most downstream in the dauer signal transduction cascade and that it directly regulates the transcription of genes necessary for dauer formation and the increased longevity of *age-1* and *daf-2* mutants.

Recently, it has been proposed that an increase in the expression of *ctl-1*, a gene which encodes a cytoplasmic catalase, could be the main cause of the increased life span of dauer constitutive (daf-c) mutants such as *daf-2* and *age-1* (Taub et al. 1999). In this publication and in others (e.g., Vanfleteren 1993), it was found that catalase expression and activity was substantially higher in dauer larvae than in other larval stages and was expressed at higher levels in old *daf-2* and *age-1* mutants than in old wild-type worms. Furthermore, it was found that a severe loss-of-function mutation in *ctl-1* completely suppresses the long life of *daf-2* and the even longer life (Larsen et al. 1995) of *daf-2 daf-12* mutants. Given these correlations, it is likely that, as the authors suggest, an altered expression of *ctl-1* is needed for the long life of dauer larvae and the daf-c mutants. However, two observations suggest that one should be cautious with the conclusion that enhanced expression of *ctl-1* causes the increase of life span of the mutants. First, attempts at transgenic overexpression of *ctl-1* have not resulted in an increased life span. Second, the *ctl-1* mutation is deleterious by itself as it shortens the average life span of animals carrying it to 76% of that of the wild-type. Further work will be necessary to say whether enhanced expression of *ctl-1* causes the increased life span of the mutants.

3
The Rate of Living

3.1
The Identification of *clk* Genes

In an effort to identify novel types of genes that regulate development, we carried out a screen for viable, maternal-effect mutations that affect the anatomy, behavior, or physiology of the worm (Fig. 5) (Hekimi et al. 1995). In *C. elegans*, embryonic development results in a small worm that is already organized exactly like the adult and that is composed of most of the post-mitotic cells that are found in the adult (Wood 1988). Very few cells are added during postembryonic development. Even though an adult worm is at least 500 times more voluminous than a freshly hatched L1 larva, this period

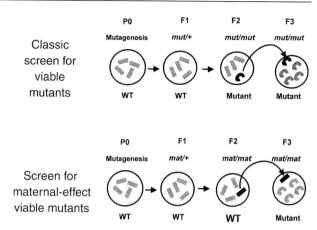

Fig. 5. Comparison between classic screening methods for finding viable recessive mutants with a visible phenotype (e.g., Brenner 1974) and the screening methods for finding maternal-effect viable recessive mutants (as carried out by Hekimi et al. 1995). Hermaphrodite worms (*gray rectangles*) are mutagenized and left to self-fertilize for two generations. In the classic screens, the F2 animals are examined visually. Animals with an mutant phenotype (*dark curved symbols*) are transferred singly onto fresh plates and left to self-fertilize. Their F3 progeny (*gray curved symbols*) are expected to show the same mutant phenotype as their F2 mother. In the maternal-effect screen, F2 animals that are phenotypically wild-type (*dark rectangle*) are transferred singly onto fresh plates and their F3 progeny is examined to identify those F2 animals which produce an entire brood of mutant progeny. The assumption at this point is that the phenotypically wild-type F2 animal that was picked was, in fact, homozygous for the mutation detected in the F3 but was maternally rescued, that is, a wild-type copy of the gene in the mother (F1) rescued the mutant phenotype in its offspring (F2). The F3 animals, however, display the mutant phenotype effect because their mother (F2) is a homozygous mutant. It is in such a screen that the *clk* genes described in the text and other mutants were identified (Hekimi et al. 1995)

of development is characterized mostly by an increase in cell size. This raised the possibility that even relatively late events of terminal differentiation could be affected by the factors contributed to the egg by the mother, and thus that mutations in genes encoding such factors would display a maternal effect.

In order to identify such genes, we designed a screen which consisted in identifying phenotypically wild-type hermaphrodites that produced only phenotypically mutant progeny and were thus likely to be carrying maternal-effect mutations (Fig. 5). In this manner, we identified 41 recessive maternal-effect mutations which fall into 24 complementation groups. Mutations in only one of these genes, *mal-2*, display a strict maternal effect. That is, heterozygous animals produced by a homozygous mutant mother are phenotypically mutant. For all other mutants isolated in our screen, the presence of a wild-type copy of the gene in a heterozygous animal was sufficient to produce a wild-type phenotype even when the mother was a homozygous mutant.

Most of the genes we identified in this screen affected the locomotory behavior or the anatomy of the worm. However, we also identified an unexpected class of genes that we have called *clk* for abnormal function of biological clocks, because mutations in these genes appear to affect many aspects of biological timing.

3.2
The *clk-1* Phenotype

Mutations in *clk-1* affect a large number of phenotypic traits that have a temporal component, including the cell cycle, embryonic development, postembryonic development, behavioral rhythms, fertility, and life span (Wong et al. 1995). Table 1 illustrates this for a number of developmental and behavioral traits for a weak (*e2519*) and a more severe (*qm30*) mutation. In *clk-1(qm30)* animals, most phenotypes we scored are slowed down approximately two-fold, on average, compared to the wild-type. This average slowdown is accompanied by a severe increase in the animal-to-animal variability of the phenotypic measures (Table 1). In addition, the cell cycle of young embryos (other cell-cycle lengths were not measured) is slowed down to the same degree as the other features (not shown). The overall slowdown of the cell cycle is due entirely to a lengthening of interphase without a concomitant slowing of mitosis. Several other phenomena are documented, including an abnormal reaction of *clk-1* mutants to changes in temperature (Wong et al. 1995).

Several observations suggest that *clk-1* plays a regulatory role in the processes that it affects. (1) In spite of their dramatic phenotype, *clk-1* mutant animals do not appear sick in any way by visual inspection. (2) The deregulation of the rate of embryogenesis can sometimes lead to embryonic

Table 1. Increased life span at all temperature for slow "rate of living" mutants

Strain	Genotype	Mean life span (days)			
		15°C	18°C	20°C	25°C
N2	+	22.0 ± 0.3	14.9 ± 0.3	16.1 ± 0.2	9.2 ± 0.3
CB4876	clk-1(e2519)	29.3 ± 0.5	18.4 ± 0.4	17.3 ± 0.4	11.6 ± 0.5
MQ130	clk-1(qm30)	31.0 ± 0.9	19.8 ± 0.5	19.3 ± 0.6	11.1 ± 0.2
MQ125	clk-2(qm37)	24.6 ± 0.8	18.7 ± 0.5	18.0 ± 0.5	11.7 ± 0.7
MQ131	clk-3(qm38)	25.7 ± 0.7	20.4 ± 0.4	19.9 ± 0.6	13.0 ± 0.8
CB4512	gro-1(e2400)	26.0 ± 0.7	20.4 ± 0.4	19.7 ± 0.6	15.6 ± 0.5

Phenotypic comparison of wild-type (N2) animals, *clk-1* mutants, and maternally rescued *clk-1 (e2519)* mutants. Developmental and behavioral phenotypes were compared among wild-type (N2) and mutants carrying either of three *clk-1* alleles; *e2519* and *qm30*. The numbers are means ± SD. In the header, *m* represents the maternal and *z* the zygotic contribution of *clk-1* activity, respectively; + and −, the wild-type and mutant form of the gene, respectively. All phenotypes were scored at 20°C.

development that is faster than that of the wild-type. (3) All phenotypes, including those measured in large adults (1000 times larger than an egg) such as behavior and life span, are fully rescued when homozygous mutants originate from a mother which carries a wild-type copy of the gene (Table 1). These observations suggest that *clk-1* affects a regulatory process that is somehow involved in setting the rate at which the organism lives its life. In particular, the observation that the maternal rescue extends to adult phenotypes and even to life span, suggests that *clk-1* can induce an epigenetic state.

3.3
Four *clk* Genes

Two additional genes, *clk-2* and *clk-3*, with phenotypes similar to *clk-1*, were identified in the same screen. A fourth gene with a Clk phenotype (*gro-1*) was originally identified on the basis of its slow growth (Hodgkin and Doniach 1997). Subsequent reappraisal of the *gro-1* mutant phenotype revealed the whole suite of features that define the Clk phenotype, including the maternal effect (Wong et al. 1995). All four genes, *clk-1*, *clk-2*, *clk-3*, and *gro-1*, have the same basic phenotype, including an increased life span, over the wild-type at all temperatures (Table 2). However, a caveat regarding the data presented in Table 1 about the *gro-1(e2400)* mutation is that it was isolated as a spontaneous mutant from a wild-type strain (PaC1) that is different from the

Table 2. Quantitative phenotypic analysis

Phenotypes	Genotypes			
	N2	$e2519\ m^+z^-$ (maternally rescued)	$e2519\ m^-z^-$	qm30
Embryonic development (hours)	13.3 ± 1.0	13.6 ± 0.8	17.1 ± 3.9	22.8 ± 5.0
Post-embryonic development (hours)	46.6 ± 3.0	50.6 ± 3.1	70.6 ± 4.8	99.2 ± 6.2
Egg production rate (eggs/hour)	6.0 ± 1.2	3.8 ± 0.9	2.7 ± 1.0	0.9 ± 0.1
Self-brood size (number of progeny)	302.4 ± 30.5	340.9 ± 77.2	191.1 ± 33.0	87.2 ± 37.2
Life span (days)	18.6 ± 5.2 (max: 27)	19.9 ± 5.3 (max: 33)	26.0 ± 9.0 (max: 45)	22.8 ± 9.8 (max: 46)
Defecation (mean of five cycle periods, in seconds)	50.8 ± 5.6	54.7 ± 8.4	69.4 ± 9.9	92.4 ± 15.0
Pumping (cycles/minute)	259.0 ± 23.7	232.3 ± 26.7	156.0 ± 29.0	170.3 ± 26.9
Swimming (cycles/minute)	120.7 ± 6.5	120.3 ± 13.2	91.7 ± 5.7	75.6 ± 4.9

clk-1, *clk-2*, *clk-3*, and *gro-1* mutants all live longer than the wild-type at all temperatures.

standard (N2) strain. As genetic background is clearly very important for the determination of life span (Johnson and Wood 1982), the use of the N2 strain as a comparison is therefore problematic. We have transferred the *gro-1* mutation into an N2 background and preliminary results do indicate that *gro-1* lengthens life span in this genetic background as well.

The observation that Clk mutants live a slow life and a long life is consistent with the "rate of living" theory of aging (Pearl 1928; Comfort 1979; Finch 1990). If life span is set by the relation between the rate of sustaining molecular injuries and their repair, then slowing down the rate of living could lead to a slower production of damage which would favor repair and thus lead to a slower accumulation of injuries. Of course, this does not tell us what the nature of these injuries is, nor does it tell us what the processes are that produce them.

3.4
The Molecular Identity of *clk-1*

We cloned *clk-1* (Ewbank et al. 1997) and showed that it encodes a phylogenetically well-conserved protein of 187-residues (Fig. 6). Mutants of the *S. cerevisiae* homologue of *clk-1*, COQ7, are incapable of growing normally on non-fermentable carbon sources. It was found that these mutants do not synthesize ubiquinone (coenzyme Q), a lipid-soluble electron carrier that is essential for respiration and consequently for growth in the absence of fermentable sources such as glucose (Marbois and Clarke 1996). Supplementation of the growth medium with CoQ partially restores growth (Jonassen et al. 1998), indicating that the mutant's absolute substrate requirement is indeed a consequence of its deficit in CoQ synthesis. Although it is possible that CLK-1 is an enzyme involved in the biosynthesis of CoQ, it does not resemble any known biosynthetic enzyme.

Coq7p has been found to be associated with the inner mitochondrial membrane in yeast (Jonassen et al. 1998). Similarly, a CLK-1::GFP fusion protein localizes to the mitochondria in worms (Fig. 7). The gene expressing the fusion protein fully rescues the phenotype of *clk-1* mutants. It should be noted, however, that there is a high level of expression in these transgenic animals and that most cells in *C. elegans* are packed with mitochondria. If 10 or 20% of CLK-1 was normally found in some other subcellular location, we might not be able to observe it. In any case, the biochemical function of the CLK-1 protein and that of its homologues is likely to be very similar because both *clk-1* and its rat homologue can rescue the phenotype of yeast *coq7* deletion mutants and restore growth on nonfermentable carbon sources (Jonassen et al. 1996; Ewbank et al. 1997).

The protein sequences of CLK-1 and its homologues can each be split and aligned to reveal a tandemly repeated domain of 82 amino acid residues

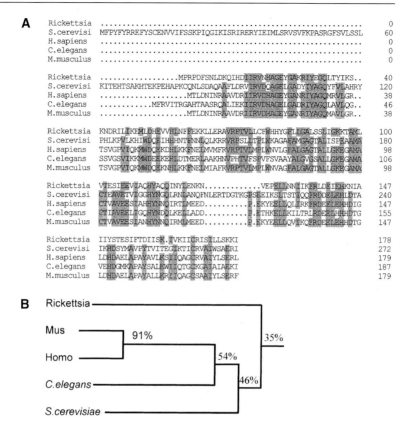

Fig. 6. A Alignment of CLK-1 amino acid sequences from a diversity of organisms. *Rickettsia prowazekii* is an intracellular parasite that is the causative agent of louse-borne typhus and is the only prokaryote in which a *clk-1* homologue has been identified so far. *R. prowazekii* is also closely related to the prokaryotic ancestor of mitochondria. Identities across all five sequences are highlighted. **B** A homology tree with the percent similarity between branches

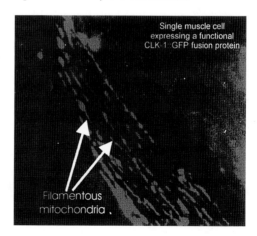

Fig. 7. Subcellular localization of a CLK-1::GFP fusion protein in the mitochondria. A single muscle cell is shown. The fluorescent mitochondria appear *white* in the figure and are roughly aligned with the contractile fibers along the main axis of the cell

(Fig. 8). When we first observed this internal homology, we concluded that the CLK-1 protein must be phylogenetically very ancient because much of the divergence between the two domains has also been conserved in yeast, worms, and rat, so that most of the overall similarity between the proteins shown in Fig. 6 is due mainly to residues that are not those that define the 82-residue domain. This implies that in a common ancestor of these very different organisms the two halves of a new gene made originally of two identical domains had time to diverge significantly in their function before the splitting of the lineage. This idea is reinforced by the recent finding of a *clk-1* homologue in the genome of the intracellular parasite, *Rickettsia prowazekii*, the causative agent of louse-borne typhus (Andersson et al. 1998). So far, among the 20 complete bacterial genomes that have been sequenced, this is the only prokaryote that has been found to harbor a *clk-1*-like sequence. *R. prowazekii* is believed to be the closest relative of the bacterium that gave rise to mitochondria. Indeed, its genome is replete with sequences that are highly homologous to eukaryotic genes that encode mitochondrial proteins (Andersson et al. 1998). However, none of the genomes of any of the free-living relatives of mitochondria and *Rickettsia* has yet been sequenced. It will be interesting to see whether they possess the *clk-1* gene. If not, it would suggest that the presence of *clk-1* is correlated with an intracellular parasitic life style.

3.5
clk-1 Mutant Mitochondria

Mitochondrial respiration is defective in yeast *coq7* mutants because of the absence of CoQ. In contrast, mitochondria purified from *clk-1* mutant worms appear to have a normal respiratory potential (Fig. 9); (Felkai et al. 1999). We used a simple colorimetric assay to test the activity of succinate cytochrome c reductase, that is, electron transfer from succinate to CoQ and from CoQ to cytochrome c. Mitochondria from *clk-1(qm30)* and *clk-1(qm51)* mutants display only a very mild reduction in respiration versus the wild-type by this assay (Fig. 9). We also tested the mitochondria with an added excess of CoQ_1 (25 µg ml^{-1}) to test whether a relative depletion of CoQ could be the primary cause of the small reduction in respiration. Adding exogenous CoQ is as effective at stimulating the enzymatic activity the wild-type than that of the mutant mitochondria. Both *clk-1(qm30)* (a partial deletion) and *clk-1(qm51)* (a splicing mutation completely abolishing the production of *clk-1* mRNA) are likely to be null mutations. Therefore, a very mild reduction in maximal respiration appears to be the null phenotype for *clk-1* (Felkai et al. 1999). That metabolic capacity is intact in *clk-1* worms is also the conclusion reached by another study which carried out various measures of metabolism in whole worms (Braeckman et al. 1999). Given this, it is very unlikely that

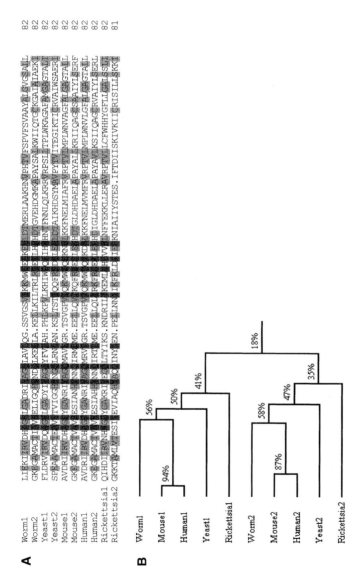

Fig. 8. **A** All CLK-1 homologues consist mainly of a tandem repeat of an 82-amino-acid-long domain. An alignment of the ten domains from five organisms is shown in the figure. Insertions and deletions within this domain in different organisms are not shown but all occur at the same point in the sequences indicated by a *dot* in the alignment. **B** A homology tree for all ten domains with the percent similarity between branches

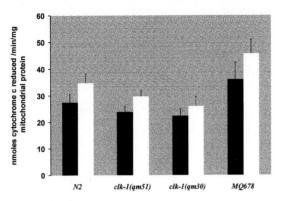

Fig. 9. The level of CLK-1 activity controls the activity of the respiratory chain over a narrow range. The activity of the mitochondrial succinate cytochrome c reductase was measured in several genotypes. This activity catalyzes the flow of electrons from succinate to CoQ and from CoQ to cytochrome c, and thus requires the CoQ cofactor. Purified mitochondria were incubated with an excess of succinate and of cytochrome c. Oxidation of cytochrome c was blocked by sodium azide and the rate of reduction of cytochrome c was monitored colorimetrically. The *black bars* represent the activity measured without added CoQ_1; the *white bars* represent the activity measured with an excess added CoQ_1 (25 µg/ml). The *error bars* represent the standard errors of the mean [N2 $n = 14$, clk-1(qm51) $n = 4$, clk-1(qm30) $n = 4$, MQ678 $n = 6$]. The genotype of the strain MQ678 is *clk-1(+); qmIs10 [clk-1::gfp]*, a strain that overexpresses CLK-1 activity. *clk-1(qm51)* and *clk-1(qm30)* are the most severe mutant alleles of *clk-1* and both lead to an average two-fold slowing of development and behavior. (Data from Felkai et al. 1999)

clk-1 encodes an enzyme of CoQ biosynthesis or that it is required for the function of such enzymes (Jonassen et al. 1998). Rather, it is likely that *clk-1* plays a regulatory role, and that the difference between yeast and worms is explained by differences in other species-specific regulatory mechanisms that also impinge on respiration and the production of CoQ.

3.6
Overexpression of CLK-1 Activity

Mutations in *clk-1* slow down the worm, but can the CLK-1 activity accelerate the worm? We found that respiration by mitochondria from worms overexpressing the *clk-1::gfp* transgene was increased, in a small but significant way, over that of wild-type mitochondria (Fig. 9). This observation suggested that overexpression might also induce a gain-of-function phenotype at the level of the whole organism. To test for gain-of-function features, we scored all life history and behavioral parameters of young transgenic animals. We found that the development, the behavior, and reproductive functions of young adult transgenic animals were not measurably different from those of the wild-type. However, it had been noted previously that the defecation cycle period, one of the physiological rates that is altered by *clk-1* mutations,

Table 3. Behavioral rate during aging in transgenic animals

Genotype	Phenotype	Defecation Day 1 (seconds)	Defecation Day 4 (seconds)	Worms with fast defecation on day 4 (% total)
clk-1(+)	Wild type	54.9 ± 0.6	106.8 ± 3.5	2.9%
clk-1(qm30)	Clk-1	81.5 ± 1.6	103.0 ± 1.7	0%
clk-1(+); qmEx133 [clk-1::gfp]	Wild type	57.1 ± 0.6	90.3 ± 3.6	24.0%
clk-1(qm30); qmEx132 [clk-1(e2519)::gfp]	Clk-1	73.7 ± 1.3	107.9 ± 4.1	0%

Overexpression of CLK-1 activity from an extrachromosomal array (*qmEx133*) partially prevents the slowing down of behavior normally occurring during aging. However, when an inactive form of CLK-1 is overexpressed [CLK-1(E2519) from array *qmEx132*] this phenomenon is not observed. Numbers are means ± SD. (Data from Felkai et al. 1999).

lengthens when worms age (Croll et al. 1977; Bolanowski et al. 1981). We tested the possibility that increased *clk-1* activity could prevent this slowdown by comparing, at different times during adult life, the defecation cycle length of wild-type animals, *clk-1* mutants, and animals overexpressing *clk-1* activity (Table 3); (Felkai et al. 1999). We found that between day 1 and day 4 of adult life there is a dramatic slowing of the defecation cycle in the wild-type, but a much lesser slowing in *clk-1* mutants. In fact, on day 4 the average period of wild-type and mutant animals is indistinguishable. This suggested that a mechanism involving *clk-1* may be responsible for the slowdown in the wild-type on day 4. Furthermore, expression of CLK-1::GFP fully prevents the slowdown in a large subset of the transgenic animals (Table 3). Indeed, 24% of the 4-day-old animals expressing the transgene are as fast on day 4 as they were on day 1, while less than 3% of the wild-type animals remain fast on day 4 (Table 3). The fact that the defecation of only a subset of the transgenic animals is as efficient on day 1 as on day 4 might be due to the existence of an additional, *clk-1*-independent, mechanism of slowdown, possibly a non-specific degradation resulting from premature aging (see below).

3.7
Acceleration of the Rate of Aging

We scored the life span of animals overexpressing transgenic CLK-1 activity and observed that they had a significantly reduced life span (Fig. 10); (Felkai et al. 1999). The short-lived transgenic animals do not appear sick or otherwise different from the wild-type, except for the absence of the behavioral decline after day 4 (see above), which suggests that their shorter life span is the result of an increase in their rate of normal aging and not the result of a novel pathology. Introduction of a mutant form of the gene in the animals had no effect on life span (Fig. 10), despite the fact that the *e2519* mutation

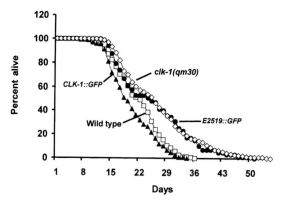

Fig. 10. CLK-1 activity controls the rate of aging in *C. elegans*. The graph shows the percentage of worms alive on a given day after having been laid during a 4-h period on day 0. The wild-type (n = 158) is shown as *filled open squares* (□), *clk-1(qm30)* (n = 199) in *open diamonds* (◇), animals from a strain that overexpresses CLK-1::GFP in a otherwise wild-type background (n = 168) in *filled triangles* (▲), and animals from a strain that overexpresses a CLK-1(E2519)::GFP (n = 87) with *filled circles* (●). Overexpression of CLK-1(+) activity (CLK-1::GFP) results in a shorter life span. However, no life span extension is observed when an almost identical protein (E2519::GFP) but with very reduced CLK-1 activity is overexpressed. The difference between the CLK-1::GFP and E2519::GFP protein consists of only a single amino acid (Ewbank et al. 1997). This indicates that it is the activity of CLK-1 is responsible for the life span shortening and not some other property of the transgene. (Data from Felkai et al. 1999)

that we used alters only one base pair and results in a single amino acid change in the CLK-1 protein (Ewbank et al. 1997). This strongly suggests that overexpression is not deleterious per se and that the effect of accelerated aging is due to the overexpression of a functional *clk-1* gene.

I have argued above that the long life of *clk-1* mutants is due to their slow life resulting in a slower production and thus accumulation of unrepaired damage. Our observations with transgenic animals overexpressing CLK-1 activity are consistent with this model. Indeed, these worms live faster than the wild-type by at least one measure (behavior), age more quickly, and have a shorter life span.

3.8
clk-1, Mitochondria and the Nucleus

The phenotype of *clk-1* mutants is quite dramatic. Almost every developmental, behavioral and reproductive rate is slowed down at least two-fold (Table 1). This is much more severe than what one would expect from the very mild defects we found in the activity of succinate cytochrome c reductase in isolated mitochondria. It is difficult to imagine that under the favorable laboratory conditions a 10% decrease in maximal respiratory output could become so limiting that it would cut developmental rates or egg production by half. Furthermore, when the food intake of worms is limited by the action of mutations in genes (*eat* genes) that are necessary for food intake, the worms have a typical "starved" appearance, being more transparent than wild-type worms and often smaller, but they show only a very mild developmental delay (Avery 1993). This is in contrast to *clk-1* mutants, that appear anatomically wild-type.

Although we have speculated that *clk-1* could be involved in the cross-talk between the mitochondria and nuclear gene expression (Felkai et al. 1999), the biochemical role of CLK-1 in the mitochondria remains unclear. However, based on our findings described in the above sections, it appears that a limitation of mitochondrial energy metabolism is not the primary defect in *clk-1* mutants.

4
Caloric Restriction

4.1
Hungry Rats

It is well known that reducing the caloric intake of rodents can significantly lengthen their mean and maximal life span (McCay et al. 1935). In fact, caloric restriction can lengthen the life span of a wide range of animals (Weindruch and Walford 1988), probably including primates (Lane et al. 1996). Not surprisingly, rodents undergoing caloric restriction display many physiological changes, including reduced body weight, body temperature, blood glucose and insulin levels, but also increases in DNA repair capabilities and decrease in oxidative damage (reviewed in Sohal and Weindruch 1996). Whether these changes correspond to a decrease in metabolic rates is, however, not yet fully settled. For example, some studies show no change in oxygen consumption per unit of lean body mass (McCarter et al. 1985; Masoro and McCarter 1991), while others show a decrease in consumption with caloric restriction (Dulloo and Girardier 1993; Ramsey et al. 1997). In any case, although it is unclear which of the changes that accompany caloric restriction are required for an extended life span (Sohal and Weindruch 1996), it is not unreasonable to suppose that a reduction in food intake should downregulate energy consumption and increase repair and thus favor survival in wait for better environmental conditions. The increase in post-reproductive life span might be a side effect of this mechanism. Although it remains unclear how exactly caloric restriction affects life span, it remains the only experimental treatment that has been shown repeatedly to significantly prolong the life of vertebrates (Rose 1991).

4.2
Hungry Worms

Lowering food intake can also extend the life span of *C. elegans* (Klass 1977, 1983). Klass clearly showed this by growing worms in liquid culture with different concentrations of bacteria. However, most aging studies in the worm have been done under standard *C. elegans* culture conditions (i.e., worms are cultured on agar plates with lawns of *E. coli*). We wanted to be able to compare the effects of caloric restriction on worm life span to those obtained by genetic means. Unfortunately, the response of many long-lived *C. elegans* strains to growth in liquid culture is unknown, and it is not even clear if all of these strains can be cultured under such conditions. Doing aging studies in liquid culture also poses difficult technical problems such as accurately maintaining and reproducing certain bacterial concentrations. We therefore turned our attention to a class of genes where mutations result in a decrease in food intake, the *eat* genes (Avery 1993).

Mutations in *eat* genes affect the function of the pharynx, a muscular organ whose rythmic contractions pump bacteria into the gut. Several *eat* mutations produce defects in the strength or coordination of the precisely timed contractions and relaxations of the pharyngeal muscles, which leads to inefficient feeding. Other mutations only reduce the rate of pumping. In addition, to their pharyngeal pumping defects, *eat* mutants have a typical "starved" appearance, that is, they are more transparent and often smaller than wild-type animals. For some *eat* mutants, there is evidence, either pharmacological, electrophysiological, genetic, or molecular, that the nervous system or muscles are affected (Davis et al. 1995; Raizen et al. 1995; Starich et al. 1996; Lee et al. 1997).

We tested the life span of a large number of *eat* mutations and found that mutations in many *eat* genes (*eat-1*, *eat-2*, *eat-3*, *eat-6*, *eat-13* and *eat-18*) significantly extend life span (Table 4). All four tested alleles of *eat-2* and *eat-6*, as well as both alleles of *eat-1* and *eat-18*, significantly increase life span (Table 4). Of all the *eat* genes tested the strongest effect was seen with *eat-2* mutants, which can live over 50% longer than the wild-type (Table 4). The life-span extension of *eat-2* is comparable to other previously characterized long-lived mutants, such as *clk-1(e2519)* and *daf-2(e1370)*, and is also of similar magnitude to the effect of caloric restriction on the life span of mammals (Weindruch et al. 1986). *eat-2* mutants have mechanically normal pumping but a very slow pumping rate. We find that the *eat-2* alleles with the slowest pumping rates, *eat-2(ad1113)* and *eat-2(ad1116)*, live longer than a clearly weaker allele, *eat-2(ad453)* (Raizen et al. 1995).

The mutations *eat-5(ad464)*, *eat-10(ad606)*, and *eat-7(ad450)* do not affect life span (Table 4). It is unclear why they do not, but it is possible that mutations in these genes lead to too weak a feeding defect to affect life span. Alternatively, these mutations might produce deleterious pleiotropic effects

Table 4. Life spans of *eat* and *unc* mutants

Genes with eating defects	Percent increase over wild type	Genes without eating defects	Percent increase over wild type
eat-1(ad427)	+33	unc-1(e538)	−5
eat-1(e2343)	+11	unc-1(719)	+3
eat-2(ad453)	+36	unc-4(e719)	+8
eat-2(ad464)	+29	unc-6(e68)	+2
eat-2(ad1113)	+46	unc-7(e5)	+1
eat-2(ad1116)	+57	unc-7(wd7)	−7
eat-3(ad426)	+11	unc-9(e101)	+27
eat-5(ad464)	−6	unc-9(nr450)	−7
eat-6(ad792)	+36	unc-9(hs6)	−11
eat-6(ad467)	+37	unc-9(ec27)	−5
eat-6(ad601)	+15	unc-24(e138)	+1
eat-6(ad997)	+19	unc-24(448)	+1
eat-7(ad450)	−35	unc-24(e927)	0
eat-10(ad606)	+8	unc-24(e1172)	+4
eat-13(ad522)	+31	unc-25(e156)	+26
eat-18(ad820sd)	+38	unc-25(e591)	+2
eat-18(ad1110)	+15	unc-25(e891)	+4
unc-26(e205)	+44	unc-25(e265)	+5
unc-26(e345)	+47	unc-30(e191)	−7
unc-26(e1196)	+37	unc-30(e318)	−22
unc-26(m2)	+35	unc-46(e177)	0
		unc-47(e367)	−5
		unc-49(e382)	+7
		unc-79(e1030)	−4
		unc-79(qm12)	−1
		unc-79(qm14)	+10
		unc-79(e1068)	+11
		unc-80(qm2)	+7
		unc-80(qm3)	+1
		unc-80(qm9)	−2
		unc-80(e1272)	+11

Decreased food intake but not abnormal locomotion increases worm life span. We tested 21 mutations that reduce food intake. These mutations are in 10 distinct genes: 9 *eat* genes that do not affect locomotion and one *unc* gene (*unc-26*) that affect locomotion as well as food intake. All strains except 3 (boxed) have a significantly increased life span. As a control we tested 32 mutations in 13 genes that are required for locomotion. All of these strains except 2 do not have an extended life span. However, the long life of animals from the long-lived strains [carrying *unc-9(e101)* and *unc-25(e156)*] is probably not due to the *unc* mutations themselves but to linked background mutations that prolong life span, as strains carrying other mutations in the same genes (boxed) do not have prolonged life spans.

which mask any positive effects on life span conferred by the caloric restriction.

The fact that long-lived *eat* mutants display a range of defects with different underlying molecular or anatomical causes strongly suggests that these mutants live long because of the only feature they share: restricted food intake. As a control, we tested whether defects in muscles and nerves that do not lead to abnormal pharyngeal pumping could also increase life span. For this we examined a number of uncoordinated (*unc*) mutants that display movement defects due to abnormalities in the function or development of the nervous system and/or the body wall muscles (Chalfie and White 1988). That the *unc* and *eat* mutations can produce similar defects is borne out by the observation that *eat-8* and *eat-11* are also Unc while *unc-2, -10, -11, -17, -18, -26, -32, -36, -37, -57, -75,* and *-104* mutants are also Eat (Avery 1993). We tested the effect of mutations in 15 *unc* genes on life span and found that mutations in 14 of these genes (*unc-1, -4, -6, -7, -9, -24, -25, -29, -30, -46, -47, -49, -79,* and *-80*) do not lengthen life span (Table 4). However, some *unc* strains appear to contain background mutations that can extend life span. For example, the strains containing *unc-9(e101)* and the strain containing *unc-25(e156)* appear to also contain life-extending background mutations that are linked to the *unc* mutation as they could not be removed by backcrosses (Table 4). However, strains with other alleles of these genes do not affect life span, suggesting very strongly that the observed increases in life span are not properties of the genes (*unc-9* or *unc-25*) themselves, but are properties of the strains that contain them.

Of the 15 *unc* genes that we examined, only mutations in *unc-26* clearly lengthen life span. Indeed, all four alleles significantly lengthen life span. However, as *unc-26* mutants are known to have a starved appearance and a feeding defect (Avery 1993), the long life of these mutants is probably also due to caloric restriction. Thus, mutations in 14/14 tested *unc* genes that do not affect pharyngeal pumping do not lengthen life span, while mutations in 7/10 genes that disrupt normal pharyngeal pumping do lengthen life span. This indicates that *eat* genes do indeed lengthen life span by caloric restriction.

5
How Many Different Mechanisms?

5.1
An Answer from Genetic Interactions

Above, evidence was presented that mutations that alter three distinct aspects of worm biology (dormancy, physiological rates, and food intake) can lengthen life span. As described, mutants from the three different categories have very different phenotypes but all are expected to affect, directly or indirectly, many aspects of cell physiology. However, at the present time, for any one mutant, it is unclear which and how many of the physiological changes are, in fact, responsible for the observed life-span extension. One way to answer this question is to ask whether the mutations in the three pathways affect aging in the same way. If any two mutations do act on life span by the same mechanism, one could focus research on what they have in common.

One way to gain more insight is to construct and analyze double mutants. This approach has been taken, for example, to show that *daf-2* and *age-1* affect aging in the same way (Dorman et al. 1995), as *daf-2 age-1* double mutants live no longer than mutants that carry only a *daf-2* or *age-1* mutation. We now know that these two genes do function in the same signal transduction pathway, where they interact very closely. On the other hand, one would imagine that if two mutations were increasing life span for different reasons, then their effects on animals could be additive (i.e., the double mutants would have a very long life). In addition, one can use *daf-16* to examine whether any mutation functions in ways similar to those of *daf-2* or *age-1* as *daf-16* mutations suppress all phenotypes of *daf-2* and *age-1* including long life. If *daf-16* were able to suppress the long life of other genes, it might suggest that their action on life span involves a mechanism similar to that of the dauer-constitutive genes.

5.2
Rate of Living and Dormancy

We compared the life span of *daf-2*; *clk-1* double mutants to that of the wild-type, *daf-2*, and *clk-1* mutants (Fig. 11). An experimental design was used (temperature shift from 18 to 25°C after the third larval molt) that favors a long life for *daf-2* mutants. Under these conditions the life span of *clk-1* was not significantly increased. However, the double mutants lived almost 6 times longer than the wild-type (Fig. 11); (Lakowski and Hekimi 1996). The worms were still producing progeny even long after all wild-type, *clk-1*, and *daf-2* worms had died. Clearly, this is an effect that goes far beyond additivity. How can we interpret this synergism? As stated above, *clk* mutants might live long because they live slowly and thus produce damage more slowly. On the other hand, *daf* mutants might live long because they are more stress-resistant (better shielding and repair) or because they have better quality control mechanisms to cope with damage. In the double mutant, less damage is produced, it is dealt with better, and it takes a very long time before the animals accumulate enough damage to kill them. In fact, the double mutants live so long that one could wonder whether it is still the same damage that kills them as that which kills wild-type worms. For example, the worm's external cuticle is constituted mostly of specialized collagens that are not being renewed after the adult molt. Any damage these proteins would sustain over time is likely to originate from environmental causes, such as UV or the action of molecular oxygen. Clearly, the structure of these collagens has not evolved to resist environmental insults for as long as they need to be functioning in the double mutants. It is possible, therefore, that double mutants no longer die from the same causes that kill wild-type animals, but from novel causes that arise from their prolonged life.

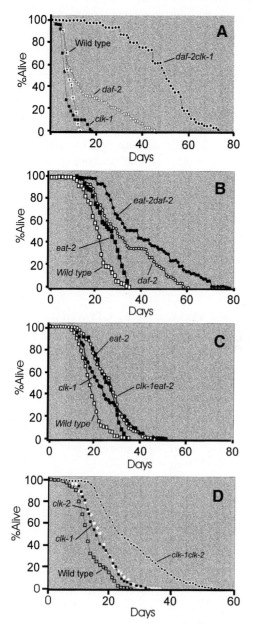

Fig. 11A–D. Death curves of *C. elegans* aging mutants and double mutant combinations. **A** *daf-2 clk-1* double mutants live much longer than the single mutants. The experiment was carried out under conditions optimal for increased life span of the temperature-sensitive *daf-2* mutants. Animals are kept at 15°C until the fourth larval stage to prevent them from entering the dauer larval stage and they are then shifted to 25°C for the rest of their life span. Under these conditions *clk-1* animals do not live longer than the wild-type. **B** *eat-2 daf-2* double mutants live longer than mutants for either of the two genes. **C** *clk-1 eat-2* double mutants do not live longer than mutants for either of the two genes. **D** *clk-1 clk-2* double mutants live substantially longer than mutants for either of the two genes. (After data in Lakowski and Hekimi 1996, 1998)

We have also examined the life span of *daf-16*; *clk-1* (Fig. 12) and *daf-16*; *clk-3* (Lakowski and Hekimi 1996), and found that *daf-16* does not suppress the long life of the *clk* genes or any other *clk-1* phenotype. Studies on the effects of *daf-16* on *clk-2* and *gro-1* are in progress. These results suggest that

Fig. 12A,B. Death curves of *C. elegans* mutants and double mutant combinations. **A** *daf-16(m26)* mutants live slightly shorter than wild-type worms. *daf-16; eat-2* mutants live slightly shorter than *eat-2* but significantly longer than the wild-type and *daf-16* mutants, indicating that *daf-16* does not suppress the long life produced by eat-2 mutations. **B** *daf-16(m26)* mutants live slightly shorter than wild-type worms. *daf-16; clk-1(qm30)* mutants live slightly shorter than *clk-1* but significantly longer than the wild-type and *daf-16* mutants, indicating that *daf-16* does not suppress the long life produced by *clk-1* mutations. (After data in Lakowksi and Hekimi 1998)

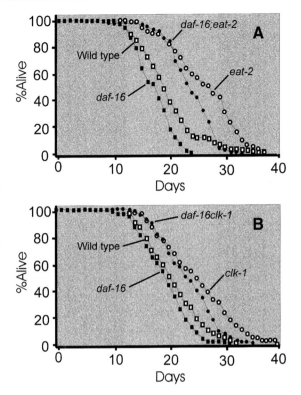

the action of the *clk* genes and the dauer-formation genes on life span are distinct. Of course, this does not strictly mean that, for aging, *clk* genes could not act downstream of *daf-16*. However, except for aging, there is very little overlap in phenotype between *daf* and *clk* genes and these phenotypes are fully additive: *daf-2clk-1* double mutants are dauer-constitutive under non-permissive conditions, they are dark as adults, and they are living much slower than *daf-2* mutants. This implies that, if *clk-1* is downstream of *daf-16* for aging, the way in which *clk-1* affects aging would have to be independent of its other functions because no state of *daf-2* or *daf-16* can mimic the other phenotypes of *clk-1* mutations.

As described above, a mutation in the gene *ctl-1* that encodes a cytosolic catalase suppresses the long life of animals carrying daf-c mutations (Taub et al. 1999). The *ctl-1* mutation also shortens the life span of *clk-1(e2519)* mutants. Thus, *ctl-1* could be seen as a possible link between these genes. However, the degree to which the *ctl-1* mutation shortens the life span of *clk-1* mutants (73%) is similar to the degree to which it shortens that of the wild-type (76%). The effect of *ctl-1* could therefore be called subtractive rather than suppressive and does not imply that *ctl-1* acts downstream of *clk-1* or that an increase in *ctl-1* activity is the basis of the long life of *clk-1* mutants.

5.3
Dormancy and Caloric Restriction

eat-2; *daf-2* double mutants were also found to live longer than either *daf-2* or *eat-2* mutants (Fig. 11), and *daf-16* does not suppress the long life of *eat-2* (Fig. 12). This indicates that, by these criteria, the genes in the dauer formation pathway extend life span by a mechanism that is distinct from that of caloric restriction. Indeed, *eat-2* cannot possibly be considered to be downstream of *daf-16*, as *eat-2* produces its effect by affecting the mechanics of the pharynx. Some observers considered this to be a surprising result (Wood 1998), because of the molecular nature of the dauer pathway genes (DAF-2 resembling an insulin-like growth factor receptors and AGE-1 being an IP3 kinase). However, signal transduction pathways can often be recruited to regulate very different downstream processes in different cells or in different animals (Brunet et al. 1999; Tan and Kim 1999).

5.4
Rate of Living and Caloric Restriction

In contrast to the previously described cases, we found the *eat-2*; *clk-1* double mutants did not live significantly longer than animals carrying only one of the component mutations (Fig. 12), suggesting that these two processes affect life span by a similar process. It should be noted, however, that *clk* and *eat* mutants have very different phenotypes and that both phenotypes can be clearly observed in double mutants. They appear starved and very active, as is characteristic of *eat-2* mutants, and have the slow development of *clk-1* mutants. This suggests that, if indeed these two types of mutants live long for the same reason, it is due to an underlying process that we do not directly observe as a phenotype. This underlying process could be some particular aspect of the function of mitochondria, where *clk-1* is localized and whose function is believed to be affected by caloric restriction. However, one should also exercise some caution in interpreting the absence of additivity, as it is also possible that the life span of the double mutants is limited by undetected deleterious effects due to the superposition of the two sets of phenotypes.

5.5
Additivity of *clk* Genes

Above, we have looked at the life span of animals carrying mutations that produce very different phenotypes and affect very different processes. Finding additivity for life span in these cases is not really unexpected. It points to the fact that life span is determined by more than one process. On the other hand, we (Lakowski and Hekimi 1996) and others (Larsen et al.

1995) have also observed additivity when two mutations affecting the same process are combined. For example, *clk-1clk-2* (Fig. 11) and *clk-3clk-2* and *clk-3clk-1* double mutants all live substantially longer than the mutants carrying only one of the component mutations. According to the logic used above, the additivity of effect of *clk* genes would mean that they affect life span by different mechanisms. However, I have also argued that the increase in life span in the *clk* mutants is an indirect consequence of their slow rate of living. *clk* double mutants live much slower than single mutants (Lakowski and Hekimi 1996) and their life is longer accordingly. This is in contrast to the other double mutants described above, where the individual mutations produce very different phenotypes that can still be observed independently in the double mutants and that, except for aging, they do not reinforce each other. The additive phenotype of the *clk* double mutants tells us that the *clk* genes do not function in a linear cascade of gene interactions and that, although they affect the rate of living by distinct mechanisms, they extend life span for the same reason, that is, a slow rate of living.

5.6
Common Grounds: Metabolic Rates and the Germline

Above, I have argued from the results of analyzing the phenotype of double mutant that there appear to be two distinct ways by which life span can be extended in worms. However, it is possible that reduced metabolic rate could be the most proximal cause of the slow aging of all long-lived strains (see the chapter by Sohal in this Volume). However, this would not be incompatible with our findings that *daf-16* does not suppress the long life of *clk-1* and *eat-2* mutants as there is no reason to suspect that the effects of *clk-1* and *eat-2* on metabolic rates require *daf-2*, *age-1*, or any of the genes that determine dauer formation. How exactly low metabolic rates extend life remains an open question.

Another common phenotype of all the long-lived mutants is the slow production of offspring (Wong et al. 1995; Lakowski and Hekimi 1996; Lakowski and Hekimi unpubl. ob.). All long-lived strains appear to produce offspring at a slower rate and, consequently, often for a longer time than the wild-type. The germline is maybe the largest tissue in adult hermaphrodites, and the production of oocytes is likely to represent the largest metabolic strain on the worms. Furthermore, Hsin and Kenyon (1999) have recently shown that the experimental ablation of the germline can lengthen worm life span and that this effect requires *daf-16* and is also affected by other *daf* genes. Although interpretation of the involvement of *daf* genes in the effect of germline ablation is difficult, the observations of Hsin and Kenyon underscore the importance of the metabolic and biosynthetic output of the organism in determining its life span.

6
A Unifying Hypothesis

Taken together, the findings about the different life-lengthening mutations in
C. elegans provide a framework for understanding the factors by which life
span is determined in worms (Fig. 13). Energy is produced in the mitochondria to support the process of living, in particular for reproduction and
somatic maintenance. However, energy has to be spent in accordance with
present and future availability. Through sensory inputs, the worm obtains
information about the presence of food and the density of other worms, and
thus a prediction of future food availability. These diverse types of information are transduced to the nucleus by the action of a signal transduction
cascade that includes *daf-2*, *age-1*, and *daf-16*, the latter actually acting in the
nucleus as a transcription factor. Through the action of *clk-1*, the nucleus also
obtains information about the functional state of the mitochondria, which
determines how much energy is available for immediate use. Furthermore,
how much food is assimilated (caloric intake) also gives very direct information about the food supply. The information about food and energy affects
the pattern of gene expression in the nucleus and, as a result, energy usage is
gated toward physiological states that consume relatively little energy (indicated by stippled arrows in Fig. 13) or states which consume large amounts of
energy (indicated by gray arrows in Fig. 13). When the environmental conditions that are sensed indicate that little food will be available in the future
(food is low and there is a high density of worms), the nucleus implements a
program of long term dormancy (the dauer stage). To ensure that the worms
survive this dormancy, some energy is spent to boost the repair process in
dauer larvae. Presumably, then, *daf-2* and *age-1* mutants that have not arrested as dauer larvae have a prolonged adult life because they retain the high

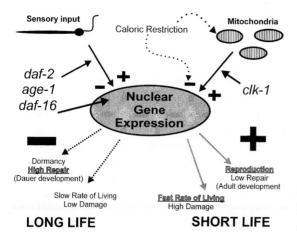

Fig. 13. A general model for the
effect of life-lengthening mutations in *C. elegans*. See text for
description

repair found in dauer larvae. When energy production by the mitochondria is inadequate (because of low temperature or environmental toxins for example) the gene expression program dictates a slow rate of living, which, in contrast to the dauer stage, can be immediately canceled and a fast rate of living resumed, and thus does not involve an investment into high repair. This slow rate of living can be seen as a short-term response to a decrease in energy availability. A slow rate of living, however, is likely to result in a low rate of damage production. In animals carrying *clk* mutations, the rate of living is continuously slow because incorrect information is being provided to the nucleus about the state of the mitochondria. These animals live long because they sustain low rates of damage. Animals that carry a mutation in a dauer stage-controlling gene and a *clk* gene live very long because they both sustain little damage and repair it efficiently.

Acknowledgments. I thank Robyn Branicky and Claire Bénard for reading the manuscript critically and in detail as well as for extensive discussions. Of course, discussion with other past and present members of my laboratory has also contributed to shape the ideas presented in this chapter. The work in my laboratory is supported by research grants from the Medical Research Council of Canada and the National Cancer Institute of Canada.

References

Anderson GL (1982) Superoxide dismutase activity in the dauer larvae of *Caenorhabditis elegans*. Can J Zool 60:288–291

Andersson SG, Zomorodipour A, Andersson JO, Sicheritz-Ponten T, Alsmark UC, Podowski RM, Naslund AK, Eriksson AS, Winkler HH, Kurland CG (1998) The genome sequence of *Rickettsia prowazekii* and the origin of mitochondria. Nature 396:133–140

Avery L (1993) The genetics of feeding in *Caenorhabditis elegans*. Genetics 133:897–917

Bolanowski MA, Russell RL, Jacobson LA (1981) Quantitative measures of aging in the nematode *Caenorhabditis elegans*. I. Population and longitudinal studies of two behavioral parameters. Mech Ageing Dev 15:279–295

Braeckman BP, Houthoofd K, De Vreese A, Vanfleteren JR (1999) Apparent uncoupling of energy production and consumption in long-lived Clk mutants of *Caenorhabditis elegans*. Curr Biol 9(9):493–496

Brenner S (1974) The genetics of *Caenorhabditis elegans*. Genetics 77:71–94

Brunet A, Bonni A, Zigmond MJ, Lin MZ, Juo P, Hu LS, Anderson MJ, Arden KC, Blenis J, Greenberg ME (1999) Akt promotes cell survival by phosphorylating and inhibiting a forkhead transcription factor. Cell 96:857–868

Chalfie M, White J (1988) The nervous system. In: The nematode *Caenorhabditis elegans*. Wood WB (ed) Cold Spring Harbor Laboratory Press, Cold Spring Harbor, New York pp 337–391

Comfort A (1979) The biology of senescence, 3rd edn. Churchill Livingstone, London

Croll NA, Smith JM, Zuckerman BM (1977) The aging process of the nematode *Caenorhabditis elegans* in bacterial and axenic culture. Exp Aging Res 3:175–189

Davis MW, Somerville D, Lee RYN, Lockery S, Avery L, Fambrough DM (1995) Mutations in the *Caenorhabditis elegans* Na,K-ATPase alpha-subunit gene, eat-6, disrupt excitable cell function. J Neurosci 15:8408–8418

Dorman JB, Albinder B, Shroyer T, Kenyon C (1995) The *age-1* and *daf-2* genes function in a common pathway to control the life span of *Caenorhabditis elegans*. Genetics 141:1399–1406

Dulloo AG, Girardier L (1993) 24-hour energy expenditure several months after weight loss in the underfed rat: evidence for a chronic increase in whole-body metabolic efficiency. Int J Obes Relat Metab Disord 17:115–123

Ewbank JJ, Barnes TM, Lakowski B, Lussier M, Bussey H, Hekimi S (1997) Structural and functional conservation of the *Caenorhabditis elegans* timing gene *clk-1*. Science 275: 980–983

Felkai S, Ewbank JJ, Lemieux J, Labbé JC, Brown GG, Hekimi S (1999) CLK-1 controls respiration behavior and aging in the nematode *Caenorhabditis elegans*. EMBO J 18(7):1783–1792

Finch CE (1990) Longevity, senescence and the genome University of Chicago Press, Chicago

Gems D, Sutton AJ, Sundermeyer ML, Albert PS, King KV, Edgley ML, Larsen PL, Riddle DL (1998) Two pleiotropic classes of *daf-2* mutation affect larval arrest, adult behavior, reproduction and longevity in *Caenorhabditis elegans*. Genetics 15:129–155

Golden JW, Riddle DL (1984) A pheromone-induced developmental switch in *Caenorhabditis elegans*: temperature-sensitive mutants reveal a wild-type temperature-dependent process. Proc Natl Acad Sci USA 81:819–823

Gottlieb S, Ruvkun G (1994) *daf-2*, *daf-16* and *daf-23*: genetically interacting genes controlling dauer formation in *Caenorhabditis elegans*. Genetics 137:107–120

Hekimi S, Boutis P, Lakowski B (1995) Viable maternal-effect mutations that affect the development of the nematode *Caenorhabditis elegans*. Genetics 141:1351–1364

Hekimi S, Lakowski B, Barnes TM, Ewbank JJ (1998) Molecular genetics of life span in *C. elegans*: how much does it teach us? Trends Genet 14:14–20

Hodgkin J, Doniach T (1997) Natural variation and copulatory plug formation in *Caenorhabditis elegans*. Genetics 146:149–164

Hsin H, Kenyon C (1999) Signals from the reproductive system regulate the life span of *C. elegans*. Nature 399:362–366

Johnson TE, Wood WB (1982) Genetic analysis of life span in *Caenorhabditis elegans*. Proc Natl Acad Sci USA 79:6603–6607

Jonassen T, Marbois BN, Kim L, Chin A, Xia YR, Lusis AJ, Clarke CF (1996) Isolation and sequencing of the rat Coq7 gene and the mapping of mouse Coq7 to chromosome 7. Arch Biochem Biophys 330:285–289

Jonassen T, Proft M, Randez-Gil F, Schultz JR, Marbois BN, Entian KD, Clarke CF (1998) Yeast Clk-1 homologue (Coq7/Cat5) is a mitochondrial protein in coenzyme Q synthesis. J Biol Chem 273:3351–3357

Kagan RM, Niewmierzycka A, Clarke S (1997) Targeted gene disruption of the *Caenorhabditis elegans* L-isoaspartyl protein repair methyltransferase impairs survival of dauer stage nematodes. Arch Biochem Biophys 348:320–328

Kenyon C, Chang J, Gensch E, Rudner A, Tabtiang R (1993) A *C. elegans* mutant that lives twice as long as wild-type. Nature 366:461–464

Kimura KD, Tissenbaum HA, Liu Y, Ruvkun G (1997) daf-2, an insulin receptor-like gene that regulates longevity and diapause in *Caenorhabditis elegans*. Science 277:942–946

Klass MR (1977) Aging in the nematode *Caenorhabditis elegans*: major biological and environmental factors influencing life span. Mech Ageing Dev 6:413–429

Klass MR (1983) A method for the isolation of longevity mutants in the nematode *Caenorhabditis elegans* and initial results. Mech Ageing Dev 22:279–286

Klass M, Hirsh D (1976) Non-ageing developmental variant of *Caenorhabditis elegans*. Nature 260:523–525

Lakowski B, Hekimi S (1996) Determination of life span in *Caenorhabditis elegans* by four clock genes. Science 272:1010–1013

Lakowski B, Hekimi S (1998) The genetics of caloric restriction in *Caenorhabditis elegans*. Proc Natl Acad Sci 95:13091–13096

Lane MA, Baer, DJ, Rumpler WV, Weindruch R, Ingram DK, Tilmont EM, Cutler RG, Roth GS (1996) Calorie restriction lowers body temperature in rhesus monkeys, consistent with a postulated anti-aging mechanism in rodents. Proc Natl Acad Sci USA 93:4159–4164

Larsen PL (1993) Aging and resistance to oxidative damage in *Caenorhabditis elegans*. Proc Natl Acad Sci 90:8905–8909

Larsen PL, Albert PS, Riddle DL (1995) Genes that regulate both development and longevity in *Caenorhabditis elegans*. Genetics 139:1567–1583

Lee RYN, Lobel L, Hengartner M, Horvitz HR, Avery L (1997) Mutations in the alpha1 subunit of an L-type voltage-activated Ca^{2+} channel cause myotonia in *Caenorhabditis elegans*. EMBO J 16:6066–6076

Lin K, Dorman JB, Rodan A, Kenyon C (1997) *daf-16*: an HNF-3/forkhead family member that can function to double the life span of *Caenorhabditis elegans*. Science 278:1319–1322

Malone EA, Inoue T, Thomas JH (1996) Genetic analysis of the roles of daf-28 and age-1 in regulating *Caenorhabditis elegans* dauer formation. Genetics 143:1193–1205

Marbois BN, Clarke CF (1996) The COQ7 gene encodes a protein in *Saccharomyces cerevisiae* necessary for ubiquinone biosynthesis. J Biol Chem 271:2995–3004

Masoro EJ, McCarter RJM (1991) Aging as a consequence of fuel utilization. Aging Clin Exp Res 3:117–128

McCarter R, Masoro EJ, Yu BP (1985) Rat muscle structure and metabolism in relation to age and food intake. Am J Physiol 248:E488–E490

McCay CM, Crowell MF, Maynard LA (1935) J Nutr 10:63–79

Morris JZ, Tissenbaum HA, Ruvkun G (1996) A phosphatidylinositol-3-OH kinase family member regulating longevity and diapause in *Caenorhabditis elegans*. Nature 382:536–539

Ogg S, Ruvkun G (1998) The *C. elegans* PTEN homolog, DAF-18, acts in the insulin receptor-like metabolic signaling pathway. Mol Cell 2:887–893

Ogg S, Paradis S, Gottlieb S, Patterson GI, Lee L, Tissenbaum HA, Ruvkun G (1997) The fork head transcription factor DAF-16 transduces insulin-like metabolic and longevity signals in *C. elegans*. Nature 389:994–999

Paradis S, Ruvkun G (1998) *Caenorhabditis elegans* Akt/PKB transduces insulin receptor-like signals from AGE-1 PI3 kinase to the DAF-16 transcription factor. Genes Dev 12:2488–2498

Pearl R (1928) The rate of living. Knopf, New York

Raizen DM, Lee RYN, Avery L (1995) Interacting genes required for pharyngeal excitation by motor neuron MC in *Caenorhabditis elegans*. Genetics 141:1365–1382

Ramsey JJ, Roecker EB, Weindruch R, Kemnitz JW (1997) Energy expenditure of adult male rhesus monkeys during the first 30 mo of dietary restriction. Am J Physiol 272:E901–E907

Rose MR (1991) Evolutionary biology of aging. Oxford University Press, New York

Riddle DL (1988) The dauer larva. In: The nematode *Caenorhabditis elegans*. Wood WB (ed) Cold Spring Harbor Laboratory Press, Cold Spring Harbor, New York pp 393–412

Riddle DL, Albert PS (1997) Genetics and environmental regulation of dauer larva development. In: *C. elegans* II. Riddle DL, Blumenthal T, Meyer BJ, Priess JR (eds). Cold Spring Harbor Laboratory Press, Cold Spring Harbor, New York pp 739–768

Sohal RS, Weindruch R (1996) Oxidative stress, caloric restriction, and aging. Science 273:59–63

Starich TA, Lee RYN, Panzarella C, Avery L, Shaw JE (1996) *eat-5* and *unc-7* represent a multigene family in *Caenorhabditis elegans* involved in cell-cell coupling. J Cell Biol 134:537–548

Tan PB, Kim SK (1999) Signaling specificity: the RTK/RAS/MAP kinase pathway in metazoans. Trends Genet 15(4):145–149

Taub J, Lau JF, Ma C, Hahn JH, Hoque R, Rothblatt J, Chalfie M (1999) A cytosolic catalase is needed to extend adult life span in *C. elegans daf-C* and *clk-1* mutants. Nature 399:162–168

Thomas JH, Birnby DA, Vowels JJ (1993) Evidence for parallel processing of sensory information controlling dauer formation in *Caenorhabditis elegans*. Genetics 134:1105–1117

Vanfleteren JR (1993) Oxidative stress and ageing in *Caenorhabditis elegans*. Biochem J 292:605–608

Weindruch RK, Walford RL (1988) The retardation of aging and disease by dietary restriction. Charles C Thomas, Springfield, Illinois

Weindruch R, Walford RL, Fligiel S, Guthrie D (1986) The retardation of aging in mice by dietary restriction: longevity, cancer, immunity and lifetime energy intake. J Nutr 116: 641–654

Wong A, Boutis P, Hekimi S (1995) Mutations in the *clk-1* gene of *Caenorhabditis elegans* affect developmental and behavioral timing. Genetics 139:1247–1259

Wood WB (1988) The nematode *Caenorhabditis elegans*. Cold Spring Harbor Laboratory Press, Cold Spring Harbor, New York pp 337–391

Wood WB (1998) Aging of *C. elegans*: mosaics and mechanisms. Cell 95:147–150

Contributions of Cell Death to Aging in *C. elegans*

Laura A. Herndon and Monica Driscoll*

1
Introduction

There is little question that cell death contributes in a significant way to development, homeostasis, and disease. The roles of cell death and cell-death genes in the process of aging, however, have not yet been clearly elaborated. The nematode *C. elegans* has proven to be a powerful genetic model system for studying both cell death and aging. Mutations affecting either cellular death or animal life span have been identified, providing the tools to evaluate genetic contributions of cell death to aging. Here, we briefly review the genetics of life span and the genetics of cell death in *C. elegans*. We discuss what is known of the contributions of apoptotic and necrotic cell death genes to aging and highlight pressing questions for more detailed evaluation of the role of cell death in aging.

2
C. elegans as Model for Analysis of Molecular Mechanisms of Aging

2.1
C. elegans as a Model System

Several features of the *C. elegans* model system are highly advantageous for the molecular and genetic dissection of aging mechanisms. This microscopic free-living nematode completes a reproductive life cycle in 2.5 days at 25°C (progressing from a fertilized embryo through four larval stages to become an egg-laying adult) and lives for a little over 2 weeks (Fig. 1). Under conditions

* *Corresponding author.*
   Department of Molecular Biology and Biochemistry, A232 Nelson Biological Laboratories, Rutgers University, 604 Allison Road, Piscataway, NJ 08854, USA

Results and Problems in Cell Differentiation, Vol. 29
Hekimi (Ed.): The Molecular Genetics of Aging
© Springer-Verlag Berlin Heidelberg 2000

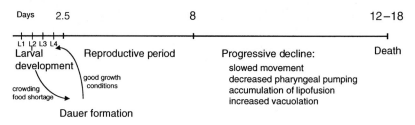

Fig. 1. Time line of the development and life of *Caenorhabditis elegans*

of starvation, overcrowding, or high temperature, larvae can enter an alternative life stage (called the dauer larvae), during which they move but do not feed (Riddle and Albert 1997). Oddly enough, dauer larvae do not "age" – dauers can survive for weeks or even months in this stage (Klass and Hirsh 1976). When food is detected, dauer larvae reenter the life cycle at the L4 larval stage and live out the rest of their "normal" life spans.

The predominant *C. elegans* sexual form is the self-fertilizing hermaphrodite. Each animal produces about 300 progeny and there is no inbreeding depression, enabling genetically identical populations to be easily maintained. Mutations affecting thousands of genes have been identified and positioned on a detailed genetic map. Until recently (Lin et al. 1998), *C. elegans* was the only multicellular organism in which it had been possible to identify single gene mutations that dramatically extend life span (Friedman and Johnson 1988a; Van Voorhies 1992; Kenyon et al. 1993; Larsen et al. 1995; Lakowski and Hekimi 1996, 1998; Yang and Wilson 1999).

Another useful feature of the *C. elegans* model system is that the development and anatomy of this tiny worm are better characterized than that of any other metazoan. The complete sequence of cell divisions that occur as the fertilized egg develops into the 959-celled adult has been recorded (Sulston and Horvitz 1977; Sulston et al. 1983). All somatic cells are postmitotic and there is no tissue regeneration, simplifying analysis of aging in nematode tissues. Serial section electron microscopy has identified the connectivity patterns for each of the 302 neurons (White et al. 1986) and provides a subcellular anatomical description of the animal from nose to tail. The *C. elegans* genome has been sequenced (*C. elegans* Sequencing Consortium 1998), providing a wealth of information for molecular study. Transgenic nematodes can be created with ease (Mello et al. 1991) and a wide variety of vectors are available for cell-specific gene expression or for generation of reporter fusions to the *E. coli* β-galactosidase gene (Fire et al. 1990) or green fluorescent protein (GFP) (Chalfie et al. 1994). Reverse genetics approaches such as gene knockouts (Zwaal et al. 1993) and double-stranded RNA-mediated gene inactivation (Fire et al. 1998) enable researchers to identify sequences with interesting homologies and to determine their loss of function phenotypes. Overall, the repertoire of experimental approaches feasible in *C. elegans* makes this organism a key model for dissection of aging mechanisms.

2.2
Characterization of Aging Nematodes

Nematodes exhibit visible changes in behavior and appearance over the course of their lives. Animals are fertile for approximately 5 days after reaching adulthood, during which time they lay approximately 300 eggs. As time progresses, animals feed (as can be measured by pharyngeal pumping), move, and defecate more slowly than their younger counterparts (Bolanowski et al. 1983; Kenyon et al. 1993; Duhon and Johnson 1995). Aging worms appear rough and lumpy, with a generally distorted morphology. One of the most striking changes in the morphology of aging animals is the dramatic appearance of vacuolated structures in the body (see further discussion below).

Scanning electron microscope studies with closely related *C. briggsae* show that old nematodes have a wrinkled cuticle when compared to young animals (Högger et al. 1977). These and other observations of the cuticle in old nematodes led to the hypothesis that old animals may lose their ability to osmoregulate through their cuticle as they age. This theory is supported by the finding that old animals are more susceptible to hypotonic shock than their younger counterparts (Zuckerman et al. 1973). Other EM studies found that old nematodes accumulate electron dense inclusions in intestinal cells (Epstein et al. 1972). These inclusions contain acid phosphatase activity (Epstein et al. 1972), suggesting that these structures might be composed of lipofuscin particles. Liposfuscin is known as an age marker since it accumulates in aging cells in most animals, including humans, and is formed from the peroxidation of lipids and proteins. The accumulation of lipofuscin particles in the gut cells of *C. elegans* was confirmed by Klass (1977).

Other changes in older nematodes include differences in their ability to tolerate various stresses. Old animals do not survive oxidative stress either from higher concentrations of oxygen (Honda et al. 1993) or consequent to exposure to oxidizing agents (Larsen 1993; Darr and Fridovich 1995) as well as younger animals. Similarly, old animals are more sensitive to thermal stress (Lithgow et al. 1994). Older nematodes lose their ability to increase their levels of superoxide dismutase (SOD) upon exposure to oxidative stress (Darr and Fridovich 1995), suggesting a mechanism for reduced survival of old animals upon exposure to stress. Studies with the long-lived mutant *age-1* revealed elevated levels of SOD and catalase as animals grow older, suggesting that the activity of antioxidant enzymes contributes to increased longevity (Larsen 1993; Vanfleteren 1993; Taub et al. 1999). *age-1* activity in wild-type animals may help to regulate SOD and catalase activity.

After oxidative stress, elevated levels of oxidized proteins result. Similarly, higher levels of protein oxidation are found in old animals (Berlett and Stadtman 1997). Protein carbonyl accumulation has been identified as a specific indicator of oxidative damage to proteins and has been found to increase with age in *C. elegans* (Adachi et al. 1998). Nematodes with different longevities appear to accumulate protein carbonyl at different rates. For example, the long-lived *age-1* and *daf-2* mutants show a slower rate of protein carbonyl accumulation while the short-lived *mev-1* mutant, which is sensitive to oxidative stress, accumulates protein carbonyl at an increased rate (Adachi et al. 1998; Yasuda et al. 1999). These results further support the link between oxidative damage and aging.

2.3
Genetics of Life Span in *C. elegans*

Several *C. elegans* genes have been found to affect life span. One set of genes, called the clock (*clk*) genes, appears to alter life span by regulating the overall rate of development (Wong et al. 1995; Lakowski and Hekimi 1996). This group of genes includes *clk-1*, *clk-2*, *clk-3*, and *gro-1*. *clk-1* encodes a highly conserved protein similar to yeast Coq7p, a metabolic regulator that may influence mitochondrial respiration (Ewbank et al. 1997). Expression of a CLK-1/GFP fusion protein revealed that CLK-1 is expressed in the mito-chondria of all somatic cells during all developmental stages (Felkai et al. 1999). In *clk-1* mutants, mitochondria are active, but their activity gradually decreases as the animals age. Although yeast mitochondria from *coq7* mu-tants have a severely defective ability to respire, respiration rate in *C. elegans* is only slightly reduced in *clk-1* mutants, suggesting that *clk-1* probably does not encode an essential enzyme of CoQ biosynthesis (Felkai et al. 1999).

Mutations in *eat* genes disrupt pharyngeal function, resulting in animals that have a lowered caloric intake. Caloric restriction has been shown to significantly lengthen life span in a wide range of animals from primates to nematodes. A survey of different *eat* mutants showed that five of eight tested had a significantly extended life span (Lakowski and Hekimi 1998). Double mutant analysis with *clk-1* showed that the double mutants live no longer than the single mutants, indicating that the *clk-1* mutations may be extending life span by a process similar to caloric restriction, possibly by reducing energy production in the worm. In contrast to the result with *clk-1*, combinations of *eat* mutants with the dauer mutants (discussed below) show an additive in-crease in life span, suggesting that food restriction and dauer mutations affect life span by different mechanisms (Lakowski and Hekimi 1998).

Another class of genes that affects the normal aging process also affects the alternative developmental pathway leading to the formation of the *C. elegans* dauer larva. Mutant alleles in the genes *age-1* (formerly also called *daf-23*) (Klass and Hirsh 1976; Friedman and Johnson 1988a; Kenyon et al. 1993; Gottlieb and Ruvkun 1994; Malone et al. 1996; Morris et al. 1996; Tissenbaum and Ruvkun 1998), *daf-2* (Kenyon et al. 1993; Tissenbaum and Ruvkun 1998), and *daf-28* (Malone et al. 1996) affect dauer development in a temperature-conditional manner, but if mutant worms are raised under conditions where they do not become dauers but gene activity is compromised, they become adults with extended life spans. Interestingly, while mutations in *daf-12* do not extend life span on their own, certain double mutant combinations of *daf-2* and *daf-12* can interact to extend life span by 250% (Kenyon et al. 1993). Some *daf-16* (Kenyon et al. 1993; Gottlieb and Ruvkun 1994; Ogg et al. 1997) and *daf-18* (Larsen et al. 1995; Mihaylova et al. 1999) alleles can suppress the life-span extensions conferred by other *age/daf* mutations, implicating these genes in the life-span extension mechanism. However, not all genes in the dauer pathway affect adult life span, and thus it is thought that the regulation of dauer formation is composed of at least two distinct pathways, only one of which is involved in life span (Gottlieb and Ruvkun 1994; Dorman et al. 1995; Morris et al. 1996; Ogg et al. 1997; Tissenbaum and Ruvkun 1998). Exactly how the *age/daf* genes interact with the *clk* genes is somewhat controversial (Lakowski and Hekimi 1996; Murakami and Johnson 1996; Braeckman et al. 1999), but in general it seems that each group defines a distinct program that affects life span via mechanisms to be determined.

By what mechanism do *age* and *daf* mutations contribute to life-span extension? Many of the life-extending mutations also confer some resistance to environmental stresses such as reactive oxygen species, UV exposure and high temperature (Larsen 1993; Vanfleteren 1993; Lithgow et al. 1995; Murakami and Johnson 1996). As mentioned above, *age-1* and *daf-2* mutants harbor increased activities of the antioxidant enzymes Cu/Zn superoxide dismutase (SOD) and catalase (Larsen 1993; Vanfleteren 1993; Vanfleteren and De Vreese 1995; Taub et al. 1999). These genes may be regulated by a signaling pathway that is involved in the modulation of stress resistance. This notion is supported by a study of nematodes carrying a mutant version of cytosolic catalase showing that *ctl-1* animals themselves have shortened life spans and suppress *age-1-*, *daf-2-*, and *clk-1*-mediated extensions of life span (Taub et al. 1999). The molecular identities of some genes affecting life extension support that these genes play a role in signaling pathways: *daf-12* encodes a member of the steroid/thyroid hormone receptor superfamily (D. Riddle, P. Larsen, W.H. Yeh, personal communication), *age-1* encodes a phosphatidylinositol-3-OH kinase (Morris et al. 1996) and *daf-2* encodes an insulin receptor-like gene (Kimura et al. 1997). *daf-16* encodes a member of the forkhead family of transcription factors (Lin et al. 1997; Ogg et al. 1997), suggesting that DAF-16 might promote expression of key longevity genes. *daf-18* encodes a homolog of the human tumor suppressor PTEN which appears to act downstream of AGE-1 to regulate the AKT-1 and AKT-2 kinases by decreasing PIP₃ levels (Ogg and Ruvkun 1998; Gil et al. 1999; Mihaylova et al. 1999; Rouault et al. 1999). Exactly how these activities function to modulate life span remains to be elaborated.

While most of the genes identified show an increase in life span due to a decrease in gene function, one gene, *tkr-1*, leads to an increase in life span and stress resistance when overexpressed. *tkr-1* encodes a putative receptor tyrosine kinase which appears to be under *daf-16* regulation (Murakami and Johnson 1998).

Animals carrying a mutation in *spe-26* also have an extended life span. Since *spe-26* mutants are defective in sperm production, initially it was thought that sperm production affected life span (Van Voorhies 1992). However, more recent studies (Gems and Riddle 1996) show that the sperm defect is separate from the longevity phenotype. Currently, it remains unclear how the *spe-26* gene, which encodes a protein similar to the actin-associated proteins kelch and scruin of *Drosophila* (Varkey et al. 1995), influences the life-span extension pathways of *C. elegans*.

3
Cell Death

Cell death plays a critical role in normal development and tissue homeo-stasis (Raff 1992; Vaux et al. 1994). Moreover, when either unscheduled implementation or inappropriate suppression of cell death occurs, patho-logical conditions can result (Thompson 1995). In landmark studies that contributed significantly to the foundation of the cell death field, Wyllie, Kerr, and colleagues suggested that diverse activating signals may converge to induce either of two morphologically distinct types of cell death, called apoptosis and necrosis (Wyllie et al. 1980; Walker et al. 1988). Apoptotic death generally occurs during normal elimination of cells during develop-ment and in cell depletion due to a broad range of stimuli, usually mild in nature. At the light microscopic level, apoptotic cell death is characterized by compaction of the dying cell. Genetic and biochemical studies have de-fined some proteins that regulate and execute apoptotic death (reviewed in Driscoll 1996; Jacobson et al. 1997; see more detailed description below). The other death type, necrosis, generally occurs in response to severe changes of physiological conditions including hypoxia, ischemia, and ex-posure to toxins, reactive oxygen metabolites, or extreme temperature. The key morphological signature of necrosis is the swelling of the dying cell. Although necrosis has been traditionally thought of as a chaotic breakdown of the cell and much less is understood of its molecular mechanism as compared to apoptosis, evidence is accumulating that this type of death may also be regulated.

3.1
Programmed Cell Death

3.1.1
Programmed Cell Death During Development in C. elegans

The nematode *Caenorhabditis elegans* has proven to be an invaluable tool for molecular and genetic dissection of the conserved apoptotic death mechanism, which is referred to as programmed cell death (PCD) in the worm (reviewed in Hengartner 1997) (Fig. 2). In the 959 somatic cells fated to live in the nematode cell lineage, the PCD program is held in check by the death-protective action of CED-9 (Hengartner et al. 1992), a functional homologue of human Bcl-2 (Vaux et al. 1992; Hengartner and Horvitz 1994). The activity of a key death-executor protein, the CED-3 caspase (Ellis and Horvitz 1986; Yuan et al. 1993; Xue et al. 1996), is positively regulated by CED-4 (related to mammalian Apaf-1: Ellis and Horvitz 1986; Shaham and Horvitz 1996; Zou et al. 1997 and FLASH: Imai et al. 1999) to initiate deaths of 131 cells fated to die during *C. elegans* development. This regulation occurs in part by protein localization to the mitochondria (Adams and Cory 1998). In cells destined to live, CED-9/Bcl-2 is situated at the mitochondrial membrane where it helps anchor the inactive CED-3 caspase. CED-4 interacts with CED-9 to release CED-3 (Chinnaiyan et al. 1997; Spector et al. 1997; Wu et al. 1997a,b; Seshagiri et al. 1998), which can oligomerize (Yang et al. 1998) and initiate a proteolytic cascade ending in cell death. PCD corpses are eliminated from the body via the phagocytotic action of neighboring cells (Robertson and

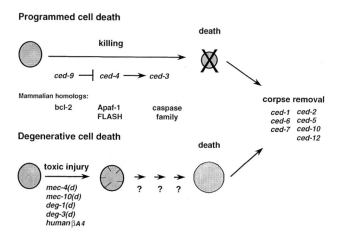

Fig. 2. Genetic pathway for programmed and degenerative cell death in *C. elegans*. Mammalian homologues are shown where identified. Degenerative cell death is regulated by a set of genes distinct from those that initiate and execute programmed cell death, though corpses are eliminated by the same genes in both pathways

Thomson 1982), a process that requires precise recognition of "dead self" from healthy living self. Corpse consumption requires the activities of at least seven genes, referred to as the engulfment *ced* genes, which function in two parallel pathways to degrade corpses (*ced-1*, *ced-6*, and *ced-7* function in one pathway; *ced-2*, *ced-5*, *ced-10*, *ced-12* function in the other) (Hedgecock et al. 1983; Ellis et al. 1991; Chung et al. 2000). A nuclease encoded or regulated by the *nuc-1* gene degrades DNA of the dead cell after it is engulfed (Sulston 1976; Hedgecock et al. 1983). An important point to note is that the basic pathway for cell death in *C. elegans* has been conserved in humans; mammalian counterparts of all cloned nematode PCD genes have been identified and implicated in human apoptosis or phagocytosis.

3.1.2
Relation of Programmed Cell Death (Apoptosis) to Aging in C. elegans

How might apoptosis contribute to the aging of an organism? One working hypothesis for the cause of aging is that cellular damage accumulates over time so that cells become increasingly dysfunctional with age. Since apoptosis is a key homeostatic mechanism that can rid the body of damaged or genetically aberrant cells, it is reasonable to propose that loss of normal controls over apoptosis could lead to the accumulation of dysfunctional cells (Wang 1997; Tomei and Umansky 1998). Another reason to hypothesize that apoptosis may influence aging concerns the intimate role of the mitochondria in both processes. Mitochondrial function/dysfunction has repeatedly been implicated in aging by the central contribution of mitochondrial action to oxidative stress (Ishii et al. 1998; Lenaz 1998) and by the correlation of mitochondrial DNA deletions with aging (Melov et al. 1995). Oxidative damage can induce apoptosis (Papa and Skulachev 1997) and mitochondrial release of cytochrome c has been shown to be an important step in the apoptotic cascade (Liu et al. 1996; Kluck et al. 1997). Likewise, opening of the mitochondrial permeability transition pore has been implicated in cell death (Kroemer 1997; Green and Reed 1998).

Mutations in the *ced-3* and *ced-4* genes disrupt all programmed cell death that occurs in *C. elegans* (Ellis and Horvitz 1986). The existence of mutants in which all apoptotic cell death is blocked has enabled a direct test of the hypothesis that programmed cell death affects aging in nematodes. Interestingly, although *ced-3* and *ced-4* mutants harbor 131 extra "undead" cells in their bodies, no major developmental or behavioral phenotypes are apparent in these mutants. *ced-3* and *ced-4* mutants do, however, have weak defects in chemotaxis that could compromise their survival in the wild (C. Bargmann, UCSF, pers. comm.). With regard to aging, it has been reported that *ced-3* and *ced-4* mutants live neither significantly shorter nor significantly longer than wild-type worms (see discussion in Kenyon 1997). Since the essential apoptosis machinery is inactivated in *ced-3* and *ced-4* mutants, it can be concluded that apoptotic-like death does not contribute in a critical way to aging in nematodes. It should be noted, however, that *C. elegans* adult somatic cells do not divide. It remains possible that the interplay between cell cycle control and apoptosis is an important factor in aging of higher organisms that continuously replenish their cells over the course of their lives.

3.2
Degenerative Cell Death

3.2.1
Necrotic-Like Cell Death in C. elegans

Inherited neurodegenerative disorders associated with necrotic-like death in *C. elegans* were identified during a screen for mutants insensitive to gentle touch (Chalfie and Sulston 1981). Many mutations in this screen affected the development or function of six touch receptor neurons. However, mutations affecting two different genes caused swelling and cell death of either the touch sensory neurons themselves (dominant mutations affecting a gene called *mec-4*; *mec*hanosensory abnormal) (Chalfie and Sulston 1981; Driscoll and Chalfie 1991; Fig. 3), or of the interneurons that send touch receptor signals to motorneurons (dominant mutations affecting a gene called *deg-1*, *deg*eneration) (Chalfie and Wolinsky 1990). *mec-4* and *deg-1* encode related ion channel subunits called degenerins that are homologous to the human epithelial amiloride-sensitive Na^+ channel and the ASIC acid-sensing channels (reviewed in Mano and Driscoll 1999). Dominant mutations in both *mec-4* (Driscoll and Chalfie 1991) and *deg-1* (García-Añoveros et al. 1995) specify the substitution of large side-chain amino acids for a conserved small residue in the vicinity of the channel pore. Such substitutions appear to increase ion influx and are thought to "lock" the channel in an open conformation. Interestingly, the initiation of *mec-4(d)*- and *deg-1(d)*-induced cell death appears reminiscent of glutamate-receptor channel mediated excitotoxic cell death that occurs in stroke (Choi 1992).

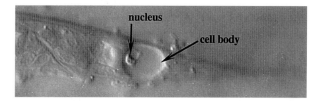

Fig. 3. Degenerative cell death of a touch receptor neuron in the tail of an L1 animal. Cell death is induced by expression of a dominant mutation in *mec-4*. Note the swelling of the cell body and the distortion of the nucleus

Necrotic-like cell death in *C. elegans* actually appears to be a general response to different "injuries". For example, a mutation that disrupts desensitization of an acetylcholine receptor channel can also induce swelling and cell death (Treinin and Chalfie 1990; Treinin et al. 1998). Another death-inducing insult is constitutive activation of a Gα_s subunit (Korswagen et al. 1997; Berger et al. 1998). Moreover, expression of human βA4 amyloid peptide in transgenic animals elicits a similar type of necrotic-like death (Link 1995). Taking into account the similar cellular responses elicited by distinct toxic insults, we have suggested that diverse initiating stimuli may elicit a common cellular response that is under genetic control (Driscoll 1996).

The necrotic cell death process induced by the hyperactive *mec-4(d)* degenerin channel is not dependent in any way upon the apoptotic cell death machinery (Ellis and Horvitz 1986; Chung et al. 2000). Mutations in *ced-9*, *ced-3*, and *ced-4* do not affect the numbers of degenerating cells, the time of degeneration onset, or the kinetics of degeneration. Thus, the necrotic-like death process is thought to occur by a distinct mechanism from apoptotic cell death (Fig. 2).

3.2.2
Neuropathology of mec-4(d)-Induced Degeneration

Examination of the ultrastructure of the touch cells from the time that the toxic *mec-4(d)* degenerin is first expressed in these cells until the cells disappear revealed a striking sequence of intracellular changes during neuro-degeneration (Hall et al. 1997). The first abnormality in a neuron expressing *mec-4(d)* is the formation of whorls of tightly wrapped electron-dense material at distinct foci at the plasma membrane. These apparent membrane whorls are internalized and appear to coalesce, growing into large membranous structures. Subsequently, large vacuoles form that distort nuclear morphology and cause chromatin clumping. Finally, cytoplasmic contents appear to lyse, often within the confines of the plasma membrane. Necrotic corpses are removed via the action of defined engulfment/phago-cytosis genes required for efficient elimination of PCD corpses (Chung et al. 2000).

Ultrastructural observations thus indicate that *mec-4(d)* expression provokes a complex sequence of intracellular events. There is some indication that similar neuropathologies associate with βA4 amyloid peptide-induced vacuolation (G. Kao, UMDNJ, pers. comm.). The nature of the event that commits the cell to destruction is not known. One possibility is that osmotic insult is the trigger. Another of many possibilities is that internalized membranes could stimulate activities of degradative lipases that could raise the intracellular concentration of toxic free radicals.

Fig. 4. Morphology of young (day 6) and old (day 16) nematodes. *Arrows* indicate the vacuolar structures that appear in older *C. elegans*

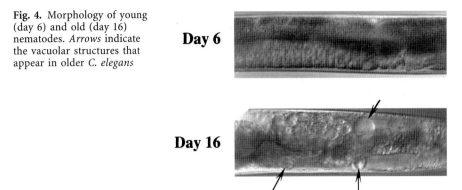

Day 6

Day 16

3.2.3
A Link Between Necrotic-Like Cell Death and Aging?

One striking feature of aging nematodes is that their bodies accumulate vacuolar figures that appear morphologically similar to the *mec-4(d)*-induced cell deaths (Fig. 4). Nothing is known of the ultrastructure of these figures in aging animals and nothing is known of their contribution to aging. Are swollen entities actually cells? Does accumulated environmental or metabolic insult delivered over time actually cause necrotic cell death? Are vacuolar figures a cause or a consequence of aging? All of these questions remain to be answered.

One line of evidence that suggests a link between necrotic-like cell death and aging comes from analysis of extragenic death suppressor mutations that block *mec-4(d)*-induced neurodegeneration (such death suppressor genes are called *des* loci for *de*generation *s*uppressors). At least one *des* gene (defined by alleles *bz29* and *bz30*) is required for toxicity of *mec-4(d)* in the touch cells, ectopically expressed *mec-4(d)* and βA4 amyloid toxicity (K. Xu and M. Driscoll, unpubl. observ.), suggesting a role for *des(bz29/30)* in a common necrotic process. To begin to address the hypothesis that the vacuolar structures which accumulate in aging animals might be related to necrotic cell deaths initiated by toxic stimuli, we tested whether *des(bz29)* and *des(bz30)* mutants had altered life spans. Indeed, we find an approximately 60% increase in life span in both independently isolated alleles of the *des(bz29/30)* gene (L. Herndon and M. Driscoll, unpubl. data), a life-span extension similar to that observed for *age-1(hx546)* mutants (Friedman and Johnson 1988b). This exciting observation has two important corollaries: first, it suggests that *des* genes may normally contribute to senescent decline in worms; and second, it raises the possibility that necrotic-like cell death may be involved in *C. elegans* aging.

4
Roles of Cell Death in *C. elegans* Aging, Future Directions

Given that cell death plays such important roles in development, homeo-stasis, and disease, it seems plausible that cell death and its regulation should influence how organisms age. The nematode model system should allow questions of how apoptotic and necrotic cell death affect senescence to be definitively addressed by assaying aging in cell death mutants. Interestingly, a first look at the relation of the normal apoptotic program to aging has established that absence of developmental PCD machinery neither slows nor accelerates aging. It remains to be determined whether other CED-3-related caspases encoded by the *C. elegans* genome (Shaham 1998) influence the aging process.

Another unanswered question regards the potential roles of the engulf-ment *ced* genes (which mediate removal of apoptotic and necrotic corpses) in senescence. One could imagine that inefficient removal of dead cells could enable toxic corpse metabolites to leak out and affect surrounding tissues, accelerating the aging process. Alternatively, since degenerative processes have been implicated in aging by the *des(bz29/30)* mutant phenotype, the disruption of the corpse degradation pathway might extend life span. It should be straightforward to evaluate life-span phenotypes of all mutants implicated in the apoptotic cell death program to provide a more thorough analysis of the relation of apoptotic cell death to aging.

Although current evidence does not favor a role for apoptotic death mechanisms in the aging of the worm, preliminary observations suggest that mechanisms operative during necrotic cell death may influence life span. The finding that mutations in one gene suppress necrotic-like cell death and also extend life span raises numerous interesting questions regarding the nature and origins of the necrotic-like figures in old worms. It will be of interest to determine how such structures are altered in *des(bz29/30)* and in the long-lived *C. elegans* mutants. Another essential line of experiments will be the evaluation of genetic interactions between *des(bz29/30)* and the *clk, eat, age/daf,* and *spe* alleles which influence life span. Finally, a clear priority should be the molecular identification of the protein encoded by *des(bz29/30)*. Prospects for addressing these questions in the immediate future appear good. Given the strong conservation of the apoptotic death mechanisms between nematodes and humans, study of genes regulating necrosis and aging in *C. elegans* may hold powerful implications for human aging and degenerative disease.

References

Adachi H, Fujiwara Y, Ishii N (1998) Oxygen is a determinant of aging in *Caenorhabditis elegans*. J Gerontol Biol Sci 53A:B240–B244

Adams JM, Cory S (1998) The Bcl-2 protein family: arbiters of cell survival. Science 281:1322–1326

Berger AJ, Hart AC, Kaplan JM (1998) Gα_s-induced neurodegeneration in *Caenorhabditis elegans*. J Neurosci 18:2871–2880

Berlett BS, Stadtman ER (1997) Protein oxidation in aging, disease, and oxidative stress. J Biol Chem 272:20313–20316

Bolanowski MA, Jacobson LA, Russell RL (1983) Quantitative measures of aging in the nematode *Caenorhabditis elegans*: II. Lysosomal hydrolases as markers of senescence. Mech Ageing Dev 21:295–319

Braeckman BP, Houthoofd K, De Vreese A, Vanfleteren JR (1999) Apparent uncoupling of energy production and consumption in long-lived Clk mutants of *Caenorhabditis elegans*. Curr Biol 9:493–496

C. elegans Sequencing Consortium (1998) Genome sequence of the nematode *C. elegans*: a platform for investigating biology. Science 282:2012–2018

Chalfie M, Sulston J (1981) Developmental genetics of the mechanosensory neurons of *Caenorhabditis elegans*. Dev Biol 82:358–370

Chalfie M, Wolinsky E (1990) The identification and suppression of inherited neurodegeneration in *Caenorhabditis elegans*. Nature 345:410–416

Chalfie M, Tu Y, Euskirchen G, Ward WW, Prasher DC (1994) Green fluorescent protein as a marker for gene expression. Science 263:802–805

Chinnaiyan A, O'Rourke K, Dixit V (1997) Interaction of CED-4 with CED-3 and CED-9: a molecular framework for cell death. Science 275:1122–1126

Choi DW (1992) Excitotoxic cell death. J Neurobiol 23:1261–1276

Chung S, Gumienny TL, Hengartner M, Driscoll M (2000) A common set of engulfment genes mediates the removal of both apoptotic and necrotic cells in *C. elegans*. (submitted)

Darr D, Fridovich I (1995) Adaptation to oxidative stress in young, but not in mature or old, *Caenorhabditis elegans*. Free Radic Biol Med 18:195–201

Dorman JB, Albinder B, Shroyer T, Kenyon C (1995) The *age-1* and *daf-2* genes function in a common pathway to control the lifespan of *Caenorhabditis elegans*. Genetics 141:1399–1406

Driscoll M (1996) Cell death in *C. elegans*: molecular insights into mechanisms conserved between nematodes and mammals. Brain Pathol 6:411–425

Driscoll M, Chalfie M (1991) The *mec-4* gene is a member of a family of *Caenorhabditis elegans* genes that can mutate to induce neuronal degeneration. Nature 349:588–593

Duhon SA, Johnson TE (1995) Movement as an index of vitality: comparing wild type and the *age-1* mutant of *Caenorhabditis elegans*. J Gerontol A Biol Sci Med Sci 50:B254–B261

Ellis HM, Horvitz HR (1986) Genetic control of programmed cell death in the nematode *Caenorhabditis elegans*. Cell 44:817–829

Ellis RE, Jacobson DM, Horvitz HR (1991) Genes required for engulfment of cell corpses during programmed cell death in *Caenorhabditis elegans*. Genetics 129:79–94

Epstein J, Himmelhoch S, Gershon D (1972) Studies on ageing in nematodes III. Electronmicroscopical studies on age-associated cellular damage. Mech Ageing Dev 1:245–255

Ewbank JJ, Barnes TM, Lakowski B, Lussier M, Bussey H, Hekimi S (1997) Structural and functional conservation of the *Caenorhabditis elegans* timing gene *clk-1*. Science 275:980–983

Felkai S, Ewbank JJ, Lemieux JJ, Labbe C, Brown GG, Hekimi S (1999) CLK-1 controls respiration, behavior and aging in the nematode *Caenorhabditis elegans*. Embo J 18:1783–1792

Fire A, Harrison SW, Dixon D (1990) A modular set of *lacZ* fusion vectors for studying gene expression in *Caenorhabditis elegans*. Gene 93:189–198

Fire A, Xu S, Montgomery MK, Kostas SA, Driver SE, Mello CC (1998) Potent and specific genetic interference by double-stranded RNA in *Caenorhabditis elegans*. Nature 391:806–811

Friedman DB, Johnson TE (1988a) A mutation in the *age-1* gene in *Caenorhabditis elegans* lengthens life and reduces hermaphrodite fertility. Genetics 118:75–86

Friedman DB, Johnson TE (1988b) Three mutants that extend both mean and maximum life span of the nematode, *Caenorhabditis elegans*, define the *age-1* gene. J Gerontol 43: B102–109

García-Añoveros J, Ma C, Chalfie M (1995) Regulation of *Caenorhabditis elegans* degenerin proteins by a putative extracellular domain. Curr Biol 5:441–448

Gems D, Riddle DL (1996) Longevity in *Caenorhabditis elegans* reduced by mating but not gamete production. Nature 379:723–725

Gil EB, Malone Link E, Liu LX, Johnson CD, Lees JA (1999) Regulation of the insulin-like developmental pathway of *Caenorhabditis elegans* by a homolog of the PTEN tumor suppressor gene. Proc Natl Acad Sci USA 96:2925–2930

Gottlieb S, Ruvkun G (1994) *daf-2, daf-16* and *daf-23*: genetically interacting genes controlling dauer formation in *Caenorhabditis elegans*. Genetics 137:107–120

Green DR, Reed JC (1998) Mitochondria and apoptosis. Science 281:1309–1312

Hall DH, Gu G, García-Añoveros J, Gong L, Chalfie M, Driscoll M (1997) Neuropathology of degenerative cell death in *C. elegans*. J Neurosci 17:1033–1045

Hedgecock EM, Sulston JE, Thomson JN (1983) Mutations affecting programmed cell deaths in the nematode *Caenorhabditis elegans*. Science 220:1277–1279

Hengartner MO (1997) Genetic control of programmed cell death and aging in the nematode *Caenorhabditis elegans*. Exp Gerontol 32:363–374

Hengartner MO, Horvitz HR (1994) *C. elegans* cell survival gene *ced-9* encodes a functional homologue of the mammalian proto-oncogene bcl-2. Cell 76:665–676

Hengartner MO, Ellis RE, Horvitz HR (1992) *Caenorhabditis elegans* gene *ced-9* protects cells from programmed death. Nature 356:494–499

Högger CH, Estey RH, Kisiel MJ, Zuckerman BM (1977) Surface scanning observations of changes in *Caenorhabditis briggsae* during aging. Nematologica 23:213–216

Honda S, Ishii N, Suzuki K, Matsuo M (1993) Oxygen-dependent perturbation of life span and aging rate in the nematode. J Gerontol 48:B57–61

Imai Y, Kimura T, Murakami A, Yajima N, Sakamaki K, Yonehara S (1999) The CED-4-homologous protein FLASH is involved in Fas-mediated activation of caspase-8 during apoptosis. Nature 398:777–785

Ishii N, Fujii M, Hartman PS, Tsuda M, Yasuda K, Senoo-Matsuda N, Yanase S, Ayusawa D, Suzuki K (1998) A mutation in succinate dehydrogenase cytochrome b causes oxidative stress and ageing in nematodes. Nature 394:694–697

Jacobson MD, Weil M, Raff MC (1997) Programmed cell death in animal development. Cell 88:347–354

Kenyon C (1997) Environmental factors and gene activities that influence life span. In: Riddle DL, Blumenthal T, Meyer BJ, Priess JR (eds) *C. elegans* II. Cold Spring Harbor Laboratory Press, Cold Spring Harbor, New York, pp 791–813

Kenyon C, Chang J, Gensch E, Rudner A, Tabtiang R (1993) A *C. elegans* mutant that lives twice as long as wild type. Nature 366:461–464

Kimura KD, Tissenbaum HA, Liu Y, Ruvkun G (1997) d*af-2*, an insulin receptor-like gene that regulates longevity and diapause in *Caenorhabditis elegans*. Science 277:942–946

Klass M, Hirsh D (1976) Non-ageing developmental variant of *Caenorhabditis elegans*. Nature 260:523–525

Klass MR (1977) Aging in the nematode *Caenorhabditis elegans*: major biological and environmental factors influencing life span. Mech Ageing Dev 6:413–429

Kluck RM, Bossy-Wetzel E, Newmeyer D (1997) The release of cytochrome c from mitochondria: a primary site for Bcl-2 regulation of apoptosis. Science 275:1132–1136

Korswagen HC, Park J-H, Ohshima Y, Plasterk RH (1997) An activating mutation in *Caenorhabditis elegans* Gα_s protein induces neural degeneration. Genes Dev 11:1493–1503

Kroemer G (1997) The proto-oncogene Bcl-2 and its role in regulation of apoptosis. Nat Med 3:614–620

Lakowski B, Hekimi S (1996) Determination of life-span in *Caenorhabditis elegans* by four clock genes. Science 272:1010–1013

Lakowski B, Hekimi S (1998) The genetics of caloric restriction in *Caenorhabditis elegans*. Proc Natl Acad Sci USA 95:13091–13096

Larsen PL (1993) Aging and resistance to oxidative damage in *Caenorhabditis elegans*. Proc Natl Acad Sci USA 90:8905–8909

Larsen PL, Albert PS, Riddle DL (1995) Genes that regulate both development and longevity in *Caenorhabditis elegans*. Genetics 139:1567–1583

Lenaz G (1998) Role of mitochondria in oxidative stress and ageing. Biochim Biophys Acta 1366:53–67

Lin K, Dorman JB, Rodan A, Kenyon C (1997) *daf-16*: an HNF-3/forkhead family member that can function to double the life-span of *Caenorhabditis elegans*. Science 278:1319–1322

Lin YJ, Seroude L, Benzer S (1998) Extended life-span and stress resistance in the *Drosophila* mutant *methuselah*. Science 282:943–946

Link CD (1995) Expression of human beta-amyloid peptide in transgenic *Caenorhabditis elegans*. Proc Natl Acad Sci USA 92:9368–9372

Lithgow GJ, White TM, Hinerfeld DA, Johnson TE (1994) Thermotolerance of a long-lived mutant of *Caenorhabditis elegans*. J Gerontol A Biol Sci Med Sci 49:B270–276

Lithgow GJ, White TM, Melov S, Johnson TE (1995) Thermotolerance and extended life-span conferred by single-gene mutations and induced by thermal stress. Proc Natl Acad Sci USA 92:7540–7544

Liu X, Kim CN, Yang J, Jemmerson R, Wang X (1996) Induction of apoptotic program in cell-free extracts: requirement for dATP and cytochrome c. Cell 86:147–157

Malone EA, Inoue T, Thomas JH (1996) Genetic analysis of the roles of *daf-28* and *age-1* in regulating *Caenorhabditis elegans* dauer formation. Genetics 143:1193–1205

Mano I, Driscoll M (1999) DEB/EnaC channels: a touchy superfamily that watches its salt. BioEssasys 21:568–578

Mello CC, Kramer JM, Stinchcomb D, Ambros V (1991) Efficient gene transfer in *C. elegans*: extrachromosomal maintenance and integration of transforming sequences. EMBO J 10:3959–3970

Melov S, Lithgow GJ, Fischer DR, Tedesco PM, Johnson TE (1995) Increased frequency of deletions in the mitochondrial genome with age of *Caenorhabditis elegans*. Nucleic Acids Res 23:1419–1425

Mihaylova VT, Borland CZ, Manjarrez LM, Stern MJ, Sun H (1999) The PTEN tumor suppressor homolog in *Caenorhabditis elegans* regulates longevity and dauer formation in an insulin receptor-like signaling pathway. Proc Natl Acad Sci USA 96:7427–7432

Morris JZ, Tissenbaum HA, Ruvkun G (1996) A phosphatidylinositol-3-OH kinase family member regulating longevity and diapause in *Caenorhabditis elegans*. Nature 382:536–539

Murakami S, Johnson TE (1996) A genetic pathway conferring life extension and resistance to UV stress in *Caenorhabditis elegans*. Genetics 143:1207–1218

Murakami S, Johnson TE (1998) Life extension and stress resistance in *Caenorhabditis elegans* modulated by the *tkr-1* gene. Curr Biol 8:1091–1094

Ogg S, Ruvkun G (1998) The *C. elegans* PTEN homolog, DAF-18, acts in the insulin receptor-like metabolic signaling pathway. Mol Cell 2:887–893

Ogg S, Paradis S, Gottlieb S, Patterson GI, Lee L, Tissenbaum HA, Ruvkun G (1997) The Fork head transcription factor DAF-16 transduces insulin-like metabolic and longevity signals in *C. elegans*. Nature 389:994–999

Papa S, Skulachev VP (1997) Reactive oxygen species, mitochondria, apoptosis and aging. Mol Cell Biochem 174:305–319

Raff MC (1992) Social controls on cell survival and cell death. Nature 356:397–400

Riddle DL, Albert PS (1997) Genetic and environmental regulation of dauer larva development. In: Riddle DL, Blumenthal T, Meyer BJ, Priess JR (eds) *C. elegans* II. Cold Spring Harbor Laboratory Press, Cold Spring Harbor, New York, pp 739–7768

Robertson AMG, Thomson JN (1982) Morphology of programmed cell death in the ventral cord of *Caenorhabditis elegans*. J Embryol Exp Morphol 67:89–100

Rouault JP, Kuwabara PE, Sinilnikova OM, Duret L, Thierry-Mieg D, Billaud M (1999) Regulation of dauer larva development in *Caenorhabditis elegans* by *daf-18*, a homologue of the tumour suppressor PTEN. Curr Biol 9:329–332

Seshagiri S, Chang WT, Miller LK (1998) Mutational analysis of *Caenorhabditis elegans* CED-4. FEBS Lett 428:71–74

Shaham S (1998) Identification of multiple *Caenorhabditis elegans* caspases and their potential roles in proteolytic cascades. J Biol Chem 273:35109–35117

Shaham S, Horvitz HR (1996) An alternatively spliced *C. elegans ced-4* RNA encodes a novel cell death inhibitor. Cell 88:201–208

Spector MS, Desnoyers S, Hoeppner D, Hengartner M (1997) Interactions between the *C. elegans* cell-death regulators CED-9 and CED-4. Nature 385:653–656

Sulston J (1976) Post-embryonic development in the ventral cord of *Caenorhabditis elegans*. Philos Trans R Soc Lond B Biol Sci 275:287–297

Sulston JE, Horvitz HR (1977) Post embryonic cell lineages of the nematode *Caenorhabditis elegans*. Dev Biol 56:110–156

Sulston JE, Schierenberg E, White JG, Thomson JN (1983) The embryonic cell lineage of the nematode *Caenorhabditis elegans*. Dev Biol 100:64–119

Taub J, Lau JF, Ma C, Hahn JH, Hoque R, Rothblatt J, Chalfie M (1999) A cytosolic catalase is needed to extend adult lifespan in *C. elegans daf-C* and *clk-1* mutants. Nature 399:162–166

Thompson CB (1995) Apoptosis in the pathogenesis and treatment of disease. Science 267:1456–1462

Tissenbaum HA, Ruvkun G (1998) An insulin-like signaling pathway affects both longevity and reproduction in *Caenorhabditis elegans*. Genetics 148:703–717

Tomei LD, Umansky SR (1998) Aging and apoptosis control. Neurol Clin 16:735–745

Treinin M, Chalfie M (1990) A mutated acetylcholine receptor subunit causes neuronal degeneration in *C. elegans*. Neuron 14:871–877

Treinin M, Gillo B, Liebman L, Chalfie M (1998) Two functionally dependent acetylcholine subunits are encoded in a single *Caenorhabditis elegans* operon. Proc Natl Acad Sci USA 95:15492–15495

Vanfleteren JR (1993) Oxidative stress and ageing in *Caenorhabditis elegans*. Biochem J 292:605–608

Vanfleteren JR, De Vreese A (1995) The gerontogenes *age-1* and *daf-2* determine metabolic rate potential in aging *Caenorhabditis elegans*. FASEB J 9:1355–1361

Van Voorhies WA (1992) Production of sperm reduces nematode lifespan. Nature 360:456–458

Varkey JP, Muhlrad PJ, Minniti AN, Do B, Ward S (1995) The *Caenorhabditis elegans spe-26* gene is necessary to form spermatids and encodes a protein similar to the actin-associated proteins kelch and scruin. Genes Dev 9:1074–1086

Vaux DL, Weissman IL, Kim SK (1992) Prevention of programmed cell death in *Cenorhabditis elegans* by human bcl-2. Science 258:1955–1957

Vaux DL, Haecker G, Strasser A (1994) An evolutionary perspective on apoptosis. Cell 76: 777–779

Walker NI, Harmon BV, Gobe GC, Kerr JF (1988) Patterns of cell death. Methods Achiev Exp Pathol 13:18–54

Wang E (1997) Regulation of apoptosis resistance and ontogeny of age-dependent diseases. Exp Gerontol 32:471–484

White JG, Southgate E, Thomson JN, Brenner S (1986) The structure of the nervous system of *Caenorhabditis elegans*. Philos Trans R Soc Lond 314:1–340

Wong A, Boutis P, Hekimi S (1995) Mutations in the *clk-1* gene of *Caenorhabditis elegans* affect developmental and behavioral timing. Genetics 139:1247–1259

Wu D, Wallen H, Nuñez G (1997a) Interaction and regulation of subcellular localization of CED-4 by CED-9. Science 275:1126–1129

Wu D, Wallen HD, Inohara N, Nunez G (1997b) Interaction and regulation of the *Caenorhabditis elegans* death protease CED-3 by CED-4 and CED-9. J Biol Chem 272:21449–21454

Wyllie AH, Kerr JF, Currie AR (1980) Cell death: the significance of apoptosis. Int Rev Cytol 68:251–306

Xue D, Shaham S, Horvitz HR (1996) The *Caenorhabditis elegans* cell-death protein CED-3 is a cysteine protease with substrate specificities similar to those of the human CPP32 protease. Genes Dev 10:1073–1083

Yang X, Chang HY, Baltimore D (1998) Essential role of CED-4 oligomerization in CED-3 activation and apoptosis. Science 281:1355–1357

Yang Y, Wilson DL (1999) Characterization of a life-extending mutation in *age-2*, a new aging gene in *Caenorhabditis elegans*. J Gerontol A Biol Sci Med Sci 54:B137–B142

Yasuda K, Adachi H, Fujiwara Y, Ishii N (1999) Protein carbonyl accumulation in aging dauer formation-defective (daf) mutants of *Caenorhabditis elegans*. J Gerontol A Biol Sci Med Sci 54:B47–53

Yuan J, Shaham S, Ledoux S, Ellis HM, Horvitz HR (1993) The *C. elegans* cell death gene *ced-3* encodes a protein similar to mammalian interleukin-1β-converting enzyme. Cell 75:641–652

Zou H, Henzel WI, Liu X, Lutschg A, Wang X (1997) Apaf-1, a human protein homologous to *C. elegans* CED-4, participates in cytochrome c-dependent activation of caspase-3. Cell 90:405–413

Zuckerman BM, Himmelhoch S, Kisiel M (1973) Fine structure changes in the cuticle of adult *Caenorhabditis briggsae* with age. Nematologica 19:109–112

Zwaal RR, Broeks A, Plasterk RH (1993) Target-selected gene inactivation in *Caenorhabditis elegans* by using a frozen transposon insertion mutant bank. Proc Natl Acad Sci USA 90:7431–7435

Stress Response and Aging in *Caenorhabditis elegans*

Gordon J. Lithgow

1
Introduction

Mutants of C. elegans that exhibit extraordinarily long life-spans are proving a rich source of information on the molecular determinates of aging rate. Such long-lived genetic variants have been subject to a range of biological, biochemical, and molecular analysis, making them an invaluable resource in the current drive to understand aging mechanisms. The mutations conferring extended life-span also confer many other phenotypes including resistance to a variety of *extrinsic* environmental stresses. This may provide a clue as to why these mutants display extended longevity. Perhaps the long-lived mutants are also resistant to *intrinsic* metabolic stresses and longevity is a consequence of an increased capacity to deal with the macromolecular damage normally caused by these stresses. However, at the time of writing there is little direct evidence for any particular mechanism for life-span extension and it is likely that different mutations extend life-span in different ways.

A mechanistic relationship between stress resistance and longevity is an emerging theme across phyla in aging research. This chapter describes the stress-resistant characteristics of the long-lived strains and attempts to assess their significance for understanding aging mechanisms. First, we should consider the ecological and evolutionary origins of aging in *C. elegans*.

2
C. elegans Life History – Life in a Stressful Environment

Soil is a ruthless and changeable environment; food may be scarce and predators abound. For soil-dwelling nematodes such as *C. elegans*, the *fungi*

The School of Biological Sciences, The University of Manchester, Oxford Road, Manchester, M13 9PT, UK

Results and Problems in Cell Differentiation, Vol. 29
Hekimi (Ed.): The Molecular Genetics of Aging
© Springer-Verlag Berlin Heidelberg 2000

imperfecti and bacterial infections (Tan et al. 1999) are a major threat to life. These fungi, with a highly effective contact adhesive to entrap nematodes and specialized body-invading hyphae, ensure that even in the absence of aging, the life-span of *C. elegans* would be short; but the soil is also a resource-rich environment. When rain and leaves fall, bacteria and slime molds abound in the nutrient-rich soil, then the free-living nematodes become the voracious predators of the bacteria and slime molds.

For the purposes of this chapter, stress is defined as events that cause cellular or neuroendocrine responses and are generally detrimental to survival. Stress can come from an extrinsic or intrinsic source. It follows that the highly changeable soil habitat that *C. elegans* inhabits is certainly stressful. This species has adapted to the rigors of the "boom-bust" soil ecology in number of ways. When environmental conditions are favorable, *C. elegans* has a fast development time and high fertility. In the presence of food, *C. elegans* has a 3-day life cycle and each self-fertilizing hermaphrodite can produce up to 300 progeny over 5 days. *C. elegans* can turn bacteria into offspring at an efficiency of 50% and the critical parameter for future reproductive success appears to be speed. Indeed, even mutations that increase total fertility place *C. elegans* at a fitness disadvantage, as they also increase generation time (Hodgkin and Barnes 1991). The combination of the speed of *C. elegans* development and its large brood size means that local food depletion, resulting in starvation, is frequent.

One of the more dramatic adaptations to starvation stress is the ability of young larvae to arrest normal development during such periods and undergo diapause. The diapause stage is achieved with the formation of a specialized larval form (the dauer larvae) which is non feeding, non reproducing, stress-resistant and, compared to a normal adult, long-lived. In the protected environment of the laboratory, the life-span of the adult hermaphrodite is around 20 days at 20°C. The dauer has a mean life-span of 60 days (Klass and Hirsh 1976).

It is the hazardous nature of the soil environment and rapid reproductive schedule that determines that *C. elegans* has a short life-span. In contrast, some nematode species enjoy less hazardous environments with lower rates of death from extrinsic causes and have longer life-spans. Some free-living species have recorded mean life spans of 5 months and parasitic nematodes have life-spans of up to 15 years (Finch 1990).

Why does a hazardous, stressful natural environment lead to a short life-span in the laboratory? One explanation comes from life history evolution theory. In this context, aging is thought to arise because of the diminishing influence of natural selection with age on the characteristics of an animal (Hamilton 1966). If we assume that genes may act at specific ages, then because of death due to extrinsic hazards, genes that act early in life will be subject to greater selective pressure than genes that act late in life (Medawar 1952). High extrinsic hazards are likely to place a high premium on the speed of development and reproduction i.e. intense selection on early acting genes. Genes that optimize development and reproduction will increase fitness and thus appear in the next generation at an increased frequency. In contrast, genes that affect post reproductive adults are unlikely to have any effect whatsoever on the next generation. It follows that aging may be the consequence of natural selection failing to select against alleles that act detrimentally at later ages. Some such genes may indeed be highly beneficial in early life but have "antagonistically pleiotropic" effects in late life (Williams 1957).

To return to mechanisms that determine life-span, such evolutionary theory suggests that the genes that determine aging are not likely to have been selected to "cause aging" but rather, the genes are likely to have primary roles in other life history traits. As we shall see, this prediction is well met in *C. elegans*.

In as much that stress, in the form of rapid environmental change, has played a significant part in molding the life history of *C. elegans*, we might also predict that genes that have a role in the major adaptations to stress should also determine longevity. This prediction is also supported by observation.

3
Longevity (Age) Mutations

The popularity of *C. elegans* as a system to investigate aging primarily stems from the existence of single gene mutations which greatly increase the mean and maximum life-span of adult worms (Age mutations). Due to the cloning of some Age genes, we know more about the molecular determinants of aging in *C. elegans* than any other species (Lithgow 1996; Hekimi et al. 1998). Age strains have been used to test various theories of aging and will undoubtedly be critical in the development of aging research in other systems; aging mechanisms uncovered in *C. elegans* will be the subject of investigations in other species.

The Age mutations are highly pleiotropic. The *clk* (for abnormal function of biological *clock*s) mutations are a good example with effects not just on life-span but on cell-cycle, embryonic and post-embryonic development, timing of the defecation cycle, swimming and food pumping (Hekimi et al. 1998).

A critical development in the *C. elegans* aging field was the realization that mutations leading to the Age phenotype also have effects on dauer formation (Kenyon et al. 1993; Larsen et al. 1995; Malone et al. 1996). Dauer larvae develop from a specialized second larval-stage (L2d) worms which themselves arise in response to a neuroendocrine signaling pathway that trans-duces information on temperature, bacteria concentration in the immediate environment, and the concentration of a pheromone produced by adult and larval worms. Consequently, dauer formation is dependent on the nutritional status of the immediate environment and the density of the surrounding worm population. Thirty or more genes control dauer formation (*daf* genes) which act in partially redundant pathways. *daf* gene mutations confer either dauer formation even under good nutritional conditions (Dauer formation-constitutive, or Daf-c) or a failure to form normal dauers under any condi-tions (Dauer formation-defective, or Daf-d). Some, but not all, Daf genes influence life-span. Mutations of either *daf-2*, *age-1*, or *daf-28* extend life-span by up to 100% (Friedman and Johnson 1988; Kenyon et al. 1993; Malone et al. 1996). Mutations of either *daf-16* or *daf-18* suppress both the Age phenotype and Daf-c phenotypes of *age-1* and *daf-2* mutations (Kenyon et al. 1993; Gottlieb and Ruvkun 1994; Dorman et al. 1995; Larsen et al. 1995; Malone et al. 1996; Tissenbaum and Ruvkun 1998). These genes all encode components of a signaling pathway resembling the vertebrate insulin sig-naling pathway: *daf-2* encodes a protein similar to insulin receptor (Kimura et al. 1997), *age-1* encodes a protein similar to phosphatidylinositol 3-OH kinase (PI3K) (Morris et al. 1996), *daf-16* encodes a fork-head transcription (Lin et al. 1997; Ogg et al. 1997), and *daf-18* encodes a protein similar to the human PTEN protein which has phosphatidylinositol 3,4,5-trisphosphate (PIP3) 3-phosphatase activity (Gil et al. 1999; Rouault et al. 1999). In sum, an insulin signaling-like pathway controls the rate of dauer formation and in-fluences life-span (Paradis and Ruvkun 1998).

The cell and tissue specificity of the action of the insulin signaling-like pathway has been investigated using a series of *daf-2* genetic mosaics (Apfeld and Kenyon 1998). It appears that *daf-2* may function in many cell types but activity in neurons and support cells is sufficient to result in normal devel-opment and normal life-span. The finding that no particular group of cells determined the developmental fate of the worm makes it likely that other signaling molecules, perhaps even a hormonal signal, act downstream of the insulin signaling pathway (Apfeld and Kenyon 1998).

Another Age mutation was originally identified as a recessive mutation affecting fertility; strains homozygous for *spe-26(hc138)* fail to produce sperm at 25°C (Varkey et al. 1996). The *spe-26(hc138)* mutation extends hermaph-rodite mean life-span by 65% (Van Voorhies 1992). The gene encodes a protein with homology to actin-binding proteins, suggesting that the SPE-26 protein is a component of the cytoskeleton (Varkey et al. 1996).

4
Aging and Stress Response

4.1
Is Aging a Stress?

Animals' responses to environmental stress include behavioral modifications, neuroendocrine changes, metabolic alterations, the activation of intracellular signaling pathways and the preferential expression of stress protein genes. Many mechanistic theories of aging have centered on the accumulation of macromolecular damage. Stress proteins generally function in preventing or repairing such damage, and consequently the relationship of stress proteins to aging is of some interest. The relationship between aging and stress response has been given considerable credence by a series of observations on the *C. elegans* Age mutants. The major finding is that mutations that lead to an extended life-span also lead to a generalized stress resistance. It follows that if an enhanced stress response counteracts aging, then aging must be equivalent to stress. Is aging simply a chronic intrinsic stress and, consequently, does an old worm equate to a stressed worm? In one respect, an aged worm does appear like a stressed worm in that they both contain elevated levels of damaged macromolecules. Such age-related changes in nematode worms have been the subject of investigation for more than 20 years (Johnson 1990) and there is evidence that many of these changes are the consequence of intrinsic or extrinsic stress. One example is the accumulation of the auto-fluorescent pigment lipofuscin, an indicator of aging in many animals including mammals. Lipofuscin is a complex mixture of lipid oxidation products and proteins that builds up in post mitotic cells. Lipofuscin was shown to accumulate with age in *C. elegans* by Klass 1977. Evidence that lipofuscin is the product of an intrinsic stress comes from a study of a strain carrying *mev-1* mutation (Hosokawa et al. 1994). This mutation is known to cause an elevated intrinsic oxidative stress (see below) and appears to accelerate the accumulation of lipofuscin.

A second common feature of an aged worm and a stressed worm is the accumulation of both covalently and conformationally altered protein. "Non covalent" conformational change probably results from oxidation events at specific amino acids, resulting in conformational isomerization and then reduction leaving the protein conformationally altered (Dulic and Gafni 1987; Yuh and Gafni 1987). Such altered protein accumulates during aging in both mammals and invertebrates. There is extensive evidence for the accumulation of conformationally modified protein with age in both *C. elegans* and the other free-living nematode, *Turbatrix aceti* (Zeelon et al. 1973; Bolla and Brot 1975; Sharma and Rothstein 1980). The accumulation of this altered protein results in a loss of specific enzyme activity (Sharma and Rothstein 1980). The reason for the accumulation of this material is unclear, but it may be due to a decrease in the rate of repair or removal of malfolded protein. A similar range of altered protein forms accumulate following stresses such as heat shock.

The idea that stress and aging are in some way linked is clearly not new and was an integral part of early proposals on the role of oxygen radicals (by products of normal metabolic activity) in aging (Harman 1956; Harman 1992). A series of recent observations in various experimental systems, but especially *C. elegans*, provides direct evidence for a mechanistic relationship between the rate of aging and stress. Most compelling is the observation that life-span extension is, almost without exception, coupled with stress resistance. We shall first consider the idea that aging is caused by the action of oxygen radicals and that the Age mutations retard this process. We shall then consider other forms of stress and what they might tell us about the action of Age mutations.

4.2
Oxidative Stress and Worm Aging

The modern free radical theory states that aging is due to reactive oxygen species (ROS) reacting with cellular components causing the accumulation of irreparable macromolecular damage (Harman 1956, 1992; Sohal 1993). Despite the large number of experiments undertaken with regard to this theory, direct evidence that ROS causes aging is scarce. However, there is considerable evidence that is at least consistent with the idea. The evidence collected from experiments with *C. elegans* includes an observed elevation in the antioxidant enzyme activity and metabolic alterations in longevous mutants, an observed failure of aged worms to induce antioxidant enzyme activities when challenged with a stress and a correlation of levels of damaged protein and DNA with age. We shall take each in turn.

Worms carrying either *age-1(hx546)* or *daf-2(e1370)* mutations were assayed for antioxidant enzyme activities Cu/Zn superoxide dismutase (SOD) and catalase (Larsen 1993; Vanfleteren 1993; Vanfleteren and De Vreese 1995). SOD dismutates the superoxide anion (O_2^-) to H_2O_2 and catalase converts H_2O_2 to H_2O and molecular oxygen. Together, these two enzymes prevent the production of the highly reactive hydroxyl radical which is probably responsible for much age-associated damage to proteins, lipids, and nucleic acids. The activities of both these enzymes are elevated in *age-1* and *daf-2* mutant worms in mid- and late life (Larsen 1993; Vanfleteren and De Vreese 1995). In addition, *age-1* mutants are resistant to *extrinsic* oxidative stress as measured by survival during exposure to hydrogen peroxide (H_2O_2) and the O_2^- generator, paraquat (Larsen 1993; Vanfleteren 1993). These results suggest that the insulin-signaling pathway negatively regulates antioxidant enzyme activity and that extended life-span results from enhanced antioxidant defences (Larsen 1993; Vanfleteren 1993). It should be noted that dauer larvae also maintain high levels of SOD (Larsen 1993; Anderson 1982) and that the elevated levels seen in the Age mutants indicate that dauer specific characters are being expressed in the long-lived adults (Kenyon et al. 1993).

Are enhanced antioxidant defences sufficient, however, to extend life-span? A series of experiments undertaken in *Drosophila* strongly suggest that they are. Transgenic *Drosophila* lines containing additional copies of the Cu/ Zn SOD and catalase genes have been generated and whilst there are some inconsistent results perhaps due to differences in experimental approaches, transgenic *Drosophila* lines which have augmented antioxidant defences also exhibit extended life-span (Orr and Sohal 1994; Parkes et al. 1998; Sun and Tower 1999).

We shall now consider the second series of observations suggesting a link between aging and oxidative stress. Aging is usually associated with frailty and an increasing susceptibility to extrinsic oxidative stress. This may be due to the effect of aging on the induction of stress response genes. Treatment of young adult worms with the pro oxidant plumbagen induces SOD activity by approximately 60% (Darr and Fridovich 1995). Hyperbaric oxygen treatment also induces SOD activity by 250%. However, when 20-day-old worms are treated with either plumbagen or hyperbaric oxygen, there is no induction in SOD activity (Darr and Fridovich 1995). The reason for this age difference is unknown, but it is probable that the failure of old animals to adapt to oxidative stress is responsible for their heightened sensitivity (Darr and Fridovich 1995).

One may speculate on what effect the failure of SOD induction would have during normal worm aging. The accumulation of oxidative damage (see below) suggests that worms do not maintain optimum antioxidant defence levels, perhaps resulting in a loss of homeostasis. Consequently, accumulated oxidative damage contributes to death. This situation is not unique to the nematode worm, as many species exhibit increasing refractivity to oxidative stress with age. In mammals, stress-activated signaling pathways are refractory to activation in aged animals with consequences for downstream stress gene regulation (Liu et al. 1996). Such studies have yet to be carried out in the worm but they have the potential of revealing critical points of failure in the gene induction and possibly demonstrate why some Age mutants can maintain higher antioxidant enzyme levels.

Other mutations that affect aging also point to a role for oxidative stress in determining life-span. Ishii and coworkers (1990) isolated a mutation in a gene called *mev-1* (methyl viogen-sensitive) that conferred sensitivity to both the superoxide generator paraquat, and to elevated oxygen tension. *mev-1* encodes a subunit of succinate dehydrogenase cytochrome b, in complex II of the mitochondrial electron transport chain. Mutations in this gene cause a reduction in complex II activity (electron transport from succinate to ubiquinone), an increase in the rate of oxidized protein accumulation (see below) and a decrease in life-span (Ishii et al. 1990; Ishii et al. 1998). One interpretation of these observations is that disruption of the electron transport chain in complex II results in an increased production of superoxide anions, leading to enhanced oxidative damage and accelerated aging.

The Clk genes may also influence aging through effects on oxidative phosphoryation. The electron transport chain is a major site of ROS production; mutations which reduce flux through the chain might also extend life-span. This is one possible interpretation of the effect of the *clk* mutations on life-span. A series of experiments by Lakowski & Hekimi (1998) examined the relationships between the Age mutations and caloric restriction extension of life-span. Caloric extension was imposed genetically by means of mutations that disrupted the pharyngeal function. Many of these mutations extend life-span. By construction of double mutant strains, Lakowski and Hekimi demonstrated that whilst insulin signaling pathway mutations exhibited synergy with caloric restriction (double mutants lived longer than either single mutant), *clk* mutants did not exhibit synergy (Lakowski and Hekimi 1998). This suggests that *clk* mutations and caloric restriction effect similar processes. The *clk-1* gene has been cloned and is the functional complement of the yeast gene CAT5. A similar gene has been found in the human genome. In *C. elegans*, the protein localizes to the mitochondria (Felkai et al. 1999).

In yeast, the CAT5 protein is involved in the synthesis of ubiquinone such that mutants have low ubiquinone levels and are respiratory deficient (Jonassen et al. 1998). Consequently, *clk-1* mutants may exhibit extended life-span as a result of reduction of flux through the electron transport chain, thus reduced oxygen radical production. However, further investigation of mitochondrial activity in *clk-1* has shown only a minor decrease in cytochrome oxidase activity (Felkai et al. 1999). Additional experiments, particularly on metabolic rates and oxygen radical production in *clk* mutants, should clarify this issue but for now it looks as if the CLK-1 protein regulates certain metabolic functions associated with the mitochondria (Felkai et al. 1999).

A very helpful addition to the study of the Age mutants in regard to oxidative stress is the development of an assay for protein carbonyl content (Adachi et al. 1998; Yasuda et al. 1999). Protein carbonyl groups ($C=O$) are formed when histidine, arginine, lysine, and proline residues are attacked by hydroxyl radicals or singlet O_2. Protein carbonyl content is commonly used as a measure, or biomarker, of aging. In animals, protein carbonyl content increases at least two fold during adult life-span (Beckman and Ames 1997). *C. elegans* is no exception. Wild-type young adults have approximately 2.5 nmoles/mg protein and this rises to nearly 5 nmoles mg^{-1} protein across the life-span (Adachi et al. 1998; Yasuda et al. 1999).

The accumulation of protein carbonyl has been studied in the Age mutants and a general inverse correlation with life-span observed (Ishii et al. 1994;

Yasuda et al. 1999). Mutation of *age-1* causes lowered levels of protein carbonyl at late ages and suggests that accumulation of this oxidative damage is slowed (Adachi et al. 1998).

Another major target for oxidative damage during aging is thought to be the DNA of the mitochondrial genome. In fact, many observers have proposed that damage to mitochondrial DNA (mtDNA) is a major cause of aging (Miquel et al. 1980; Miquel 1992). This proposal is based on the observation of an increase in the occurrence of mutated, usually deleted or rearranged, mtDNA (dmtDNA) during aging (Cortopassi and Arnheim 1990; Linnane et al. 1990; Beckman and Ames 1998). The presence of these deletions is usually correlated with a functional deficit (Corral-Debrinski et al. 1992; Hayakawa et al. 1993; Lezza et al. 1994; Melov et al. 1995). *C. elegans* has a 13.8 kb mtDNA genome encoding 12 proteins, 22 tRNA genes and 2 ribosomal RNA genes. As is the case with a variety of mammalian species, dmtDNA occurs with increasing frequency during aging of *C. elegans* despite its short life-span (Melov et al. 1994; Melov et al. 1995). These deleted forms were detected by means of a PCR reaction that amplifies up to 6.3 kb of the genome. Deletion events between the primer sites of this amplicon produce smaller PCR products. Such altered mtDNA forms accumulate, not just in wild-type strains, but also in a strain carrying the *age-1(hx546)* mutation.

Many questions are posed by these observations and by the presence of dmtDNA in aging animals in general. How well does the accumulation of dmtDNA predict life-span? If, indeed, dmtDNA results from the action of oxygen radicals, then mutations that enhance resistance to this intrinsic stress may be expected to retard the rate at which they occur. We might imagine that some of the Age mutants would have such an effect. It is not clear from the studies undertaken to date that this is the case. If accumulation of dmtDNA does predict life-span, is this because the presence of dmtDNA is detrimental and causes aging? Again, there are no studies that have addressed this question directly.

4.3
Thermotolerance and the Age Mutants

Acute temperature elevation induces a coordinate response in all organisms: the heat-shock response (Morimoto et al. 1994; Parsell and Lindquist 1994). Thermal stress leads to the accumulation of malfolded or conformationally altered protein that triggers the expression of gene families that collectively encode the heat-shock proteins (HSPs). HSP genes are expressed in response to many agents other than heat, including oxidative stress, heavy metals, radiation, and amino acid analogues. These agents all have in common the property of inducing the formation of altered protein forms. Each of these different forms of stress induces overlapping but distinct sets of HSP genes and other stress genes including antioxidant enzymes SOD and catalase (Hass and Massaro 1988; Yamashita et al. 1997; Yamashita et al. 1998).

The potential link between the well-studied hsp gene families and aging originated with the observation that when the nematodes were subjected to lethal heat stress, most Age mutations tested demonstrated increased thermotolerance (Itt) (Lithgow et al. 1995). Itt was assayed by shifting synchronously aging hermaphrodite animals from 20 to 35°C and monitoring survival. Strains carrying the mutation *age-1(hx546)* exhibit a 40% increase in survival. When worm of different ages were tested, thermotolerance was shown to decline with age but the Itt phenotype of *age-1(hx546)* can be detected at all ages (Lithgow et al. 1994).

Other Daf mutations also confer Itt including *daf-7*, *daf-4* (both mutations effecting a TGFβ signaling pathway), *daf-2* and *daf-28* mutations (Lithgow et al. 1995; GJL unpubl. data). Both *daf-2* and *daf-28* mutations lead to extended life-span and since mutations in *spe-26* (Lithgow et al. 1995), *clk-1*, and *gro-1* also confer Itt (T. Johnson, pers. comm.), it appears that Itt is necessary, but not sufficient, for extended life-span.

Could the physiological alteration that confers Itt also confer Age? A genetic dissection of pathways leading to Itt may prove useful in answering this question. In an extensive investigation of an alleleic series of *daf-2*, Gems and co-workers demonstrated a tight correlation between the effect of 16 *daf-2* mutations on life-span and the effect on thermotolerance (Gems et al. 1998). They proposed two classes of *daf-2* mutations; class 1 mutations confer Daf-c, Age and Itt and the phenotypic effects are correlated. Class 2 mutations also confer Daf-c, Age, and Itt but also one or more of the following phenotypes: reduced adult motility, abnormal adult body and gonad morphology, high levels of embryonic and L1 arrest, production of progeny late in life, and reduced brood size. The class 2 phenotypes do not correlate well with the class 1 phenotypes consistent with the idea that class 1 adult phenotypes, Age and Itt, are the result of the same physiological alteration.

A second way to address the relationship between Itt and Age is through further addressing the genetic commonality of the two phenotypes. A screen for novel mutations that confer Itt also yielded mutants that were Age, suggesting that there may be a close mechanistic relationship between the two phenotypes (Walker et al. 1998; Walker et al. 1998). Selection of novel mutations that extend life-span by screening on such a surrogate phenotype might facilitate the rapid identification of new genes (Lin et al. 1998; Walker et al. 1998).

The genetic data suggest that the DAF-2/AGE-1 insulin-like signaling pathway regulates thermotolerance. This is consistent with a study that demonstrates that when the insulin signaling pathway is inhibited by the action of a chemical inhibitor of mammalian PI3Ks (LY294002), dauer formation is induced. When young adults are treated with the drug and subjected to heat shock, survival is very significantly increased, suggesting that the insulin signaling pathway is active in adult animals and depresses the stress response. The drug also induces a small but significant increase in mean life-span (Baber et al. 1999).

Another class of genes also regulates thermotolerance in *C. elegans*. Transgenic lines maintaining extra copies of the *tkr-1* gene encoding a putative tyrosine kinase receptor were Itt, resistant to ultra violet (UV) radiation (see below) and are 65% longer-lived (Murakami and Johnson 1998). This study was the first demonstration of life-span extension by transgenesis in *C. elegans* and strongly indicates a mechanistic link between thermotolerance and life-span (Murakami and Johnson 1998).

There is very good evidence from a range of experimental systems that expression of *hsp* genes is sufficient to increase thermotolerance. Are HSPs, in part, responsible for adult thermotolerance of Age mutants? If so, are HSPs also responsible for the life-span extension? Most HSPs are molecular chaperones that interact with other cellular proteins, preventing their aggregation. Molecular chaperones have roles in the regulation of protein folding, protein translocation, the assembly of protein complexes, and protein degradation. These functions are essential in the absence of heat shock and consequently some members of HSP gene families are constitutively expressed. Molecular chaperones prevent protein aggregation by recognizing and binding to malfolded proteins and can even facilitate the refolding of malfolded protein, in an energy-dependent fashion. Since thermal stress leads to the unfolding of proteins, chaperone function is important for organismal survival following stress (Parsell and Lindquist 1994).

The *C. elegans* genome contains homologues of most molecular chaperone gene families. The best- studied family in *C. elegans* is the small heat-shock protein, HSP-16. The ubiquitous use of transcriptional promoter sequences from *hsp-16* loci for heterologous gene expression has led to wide availability of HSP-16 transcriptional reporter systems. HSP-16 is related to the mammalian α-crystallin hsp-27 (Jones et al. 1986; Candido et al. 1989) that confers thermotolerance upon over expression in cell culture (Landry et al. 1989; Rollet et al. 1992). Strains carrying Age mutations in the *age-1* gene over accumulate HSP-16, compared to wild type strains, under conditions in which the Itt phenotype is observed (G.J. Lithgow, G.A. Walker, T. White and T.E. Johnson, unpubl. data). Consequently, AGE-1 protein is likely to be a negative regulator of genes encoding HSP-16. The speculative suggestion that follows is that the over expression of *hsp* genes, such as HSP-16, in Age strains leads to a better capacity to deal with the accumulation of conformationally altered protein with age, and consequently to an extended life-span.

There is currently no direct evidence, however, that levels of molecular chaperones influence worm survival under normal conditions. However, supporting evidence come from the observation that mild non lethal heat treatment of young adult hermaphrodites significantly increases both subsequent thermotolerance and also life-span (Lithgow et al. 1995). Likewise, exposure to non lethal doses of ionizing radiation has a similar effect on life-span (Johnson and Hartman 1988). A simple interpretation of these experiments is that these mild stress induce the accumulation of stress response proteins, such as molecular chaperones, that retard aging.

4.4
UV Resistance and Aging

Age mutations also confer resistance to ultraviolet radiation (Murakami and Johnson 1996). In a study of all published Age mutations, including *spe-26* and *clk-1* mutations, young adults exhibited increased resistance to a lethal UV exposure (Uvr phenotype). UV resistance is perhaps a better predictor of life-span than thermotolerance, as Itt *daf-4* and *daf-7* alleles that exhibit normal life-spans also exhibit normal survival following UV treatment. Murakami and Johnson also demonstrated that the Uvr phenotype of *daf-2(e1370)*, *age-1(hx546)*, and *spe-26(hc138)* is suppressed by mutation of the *daf-16* gene (Murakami and Johnson 1996). UV resistance and life-span extension is also conferred by the over expression of *tkr-1* (Murakami and Johnson 1998).

The mode of toxicity of UV on worms is not known. Together with the formation of pyrimidine diamers, UV-B (290–320 nm) also can convert H_2O_2 to 2 OH•. These hydroxyl radicals are most likely responsible for the formation of 8-hydroxyguanine in DNA from UV-B-exposed mouse keatinocytes (Maccubbin et al. 1995). Consequently, UV toxicity in worms could be analogous to an oxidative stress and the resistance phenotype may arise from over expression of catalase that would lower the H_2O_2 pool.

5
Stress and Life-span Determination

In summary, there are many indications that genes that determine aging rate in *C. elegans* also affect stress resistance. This may be due to the altered expression of stress response, but also to metabolic alterations. If we take as a general mechanistic model of aging that increased likelihood of death with advancing age is a consequence of the rate of accumulation of macromolecular damage due to intrinsic metabolic stress, then one would not be surprised that genes that influence the rate of damage or repair should determine life-span. Some of the genes that influence aging are involved in a major developmental response to starvation stress, and others influence the activity and perhaps the efficiency of the oxidative phosphorylation, the origin of intrinsic oxidative stress.

We have seen that the Age mutations of *C. elegans* are highly pleiotropic and, to a large extent, share a series of stress-resistant phenotypes. What, if any, is the significance? It is also clear that many of the Age mutations cause alterations to metabolism. Is metabolic efficiency, macromolecular-damage accumulation, and stress resistance linked?

The relationship between stress and aging has been probed in other species, particularly *Drosophila*. Correlation of stress resistance and longevity has been observed in artificial selection studies on *Drosophila* species. Selection for desiccation resistance in large outbred populations led to lines which had extended longevity (Rose et al. 1992; Hoffman and Parsons 1993). As is the case with *C. elegans*, survival was enhanced in *Drosophila* by the treatment of young populations with mild thermal stress (Maynard Smith 1958a,b; Khazaeli et al. 1997). This effect on survival was even greater in strains of flies that carry extra copies of inducible *hsp-70* gene (Tatar et al. 1997), providing direct evidence that enhancing stress resistance mechanisms lowers mortality.

Single gene mutations in organisms other than *C. elegans* also provided evidence of a link to stress resistance. Mutation of the *Drosophila methuselah* (mth) gene, that encodes a guanosine triphosphate-binding protein-coupled seven-transmembrane domain receptor, confers a 35% increase in mean life-span (Lin et al. 1998). In addition, mutation of the gene confers resistance to starvation, thermal stress and the superoxide anion generator, paraquat (Lin et al. 1998). Even mutations which extend the replicative life-span of budding yeast are resistant to nutrient deprivation (Kennedy et al. 1995).

Mammals also exhibit correlation between life-span and stress resistance. Calorically restricted rodents are resistant to thermal stress probably through increased expression of hsp genes (Heydari et al. 1993). More generally cell cultures established from a wide range of mammalian species exhibit stress resistance in proportional to their normal life-spans (Kapahi et al. 1999).

Taken together with the *C. elegans* work, these observations prompt the notion of a mechanistic commonality in aging mechanisms across species (Johnson et al. 1996; Martin et al. 1996). Attempts have been made to frame a "stress hypothesis of aging" which promotes the idea that selection pressure on stress resistance characters has implications for longevity (Parsons 1993, 1995, 1996). Since there exists a large and varied literature on organismal responses to stress and on stress response proteins, one might predict rapid progress in this area of aging research with possible implications for understanding the etiology of age-related disease in humans.

References

Adachi H, Fujiwara Y, Ishii N (1998) Effects of oxygen on protein carbonyl and aging in
    Caenorhabditis elegans mutants with long (*age-1*) and short (*mev-1*) life spans. J Gerontol A
    Biol Sci Med Sci 53A : B240–B244

Anderson GL (1982) Superoxide dismutase activity in dauer larvae of *Caenorhabditis elegans*
    (Nematoda: Rhabditidae). Canadian Journal of Zoology 60 : 288–291

Apfeld J, Kenyon C (1998) Cell nonautonomy of *C. elegans daf-2* function in the regulation of
    diapause and life span. Cell 95 : 199–210

Baber P, Adamson C, Walker GA, Walker DW, Lithgow GJ (1999) PI3-kinase Inhibition Induces
    Dauer Formation, Thermotolerance and Longevity in *C. elegans*. Neurobiol Aging (in press)

Beckman KB, Ames BN (1997) Oxidants, antioxidants and aging. In: Scandalios JG (ed) Oxi-
    dative stress and the molecular biology of antioxidant defences. CSHL Press, New York, pp
    201–246

Beckman KB, Ames BN (1998) Mitochondrial aging: open questions. Ann NY Acad Sci
    854 : 118–127

Bolla R, Brot N (1975) Age dependent changes in enzymes involved in macromolecular syn-
    thesis in *Turbatrix aceti*. Arch Biochem Biophy 169 : 227–236

Candido EP, Jones D, Dixon DK, Graham RW, Russnak RH, Kay RJ (1989) Structure, organi-
    zation, expression of the 16-kDa heat shock gene family of *Caenorhabditis elegans*. Genome
    31 : 690–697

Corral-Debrinski M, Shoffner JM, Lott MT, Wallace DC (1992) Association of mitochondrial
    DNA damage with aging and coronary atherosclerotic heart disease. Mutat Res 275 : 169–180

Cortopassi GA, Arnheim N (1990) Detection of a specific mitochondrial DNA deletion in tissues
    of older humans. Nucleic Acids Res 18 : 6927–6933

Darr D, Fridovich I (1995) Adaptation to oxidative stress in young, but not in mature or old,
    Caenorhabditis elegans. Free Radic Biol Med 18 : 195–201

Dorman JB, Albinder B, Shroyer T, Kenyon C (1995) The *age-1* and *daf-2* genes function in a
    common pathway to control the lifespan of *Caenorhabditis elegans*. Genetics 141 : 1399–1406

Dulic V, Gafni A (1987) Mechanism of aging of rat muscle glyceraldehyde-3-phosphate dehy-
    drogenase studied by selective enzyme-oxidation. Mech Ageing Dev 40 : 289–306

Felkai S, Ewbank JJ, LemieuxJ, Labb, Brown GG, Hekimi S (1999) CLK-1 controls respiration,
    behavior and aging in the nematode *Caenorhabditis elegans*. EMBO J 18 : 1783–1792

Finch CE (1990) Longevity, senescence and the genome. University of Chicago Press, Chicago
    and London

Friedman DB, Johnson TE (1988) A mutation in the *age-1* gene in *Caenorhabditis elegans*
    lengthens life and reduces hermaphrodite fertility. Genetics 118 : 75–86

Friedman DB, Johnson TE (1988) Three mutants that extend both mean and maximum life span
    of the nematode, *Caenorhabditis elegans*, define the *age-1* gene. J Gerontol Biol Sci 43 : B102–
    B109

Gems D, Sutton AJ, Sundermeyer ML, Albert PS, King KV, Edgley ML, Larsen PL, Riddle DL
    (1998) Two pleiotropic classes of *daf-2* mutation affect larval arrest, adult behavior, re-
    production and longevity in *Caenorhabditis elegans*. Genetics 150 : 129–155

Gil EB, Malone LE, Liu LX, Johnson CD, Lees JA (1999) Regulation of the insulin-like devel-
    opmental pathway of *Caenorhabditis elegans* by a homolog of the PTEN tumor suppressor
    gene. Proc Natl Acad Sci USA 96 : 2925–2930

Gottlieb S, Ruvkun G (1994) *daf-2*, *daf-16* and *daf-23*: Genetically interacting genes controlling
    dauer formation in *Caenorhabditis elegans*. Genetics 137 : 107–120

Hamilton WD (1966) The moulding of senescence by natural selection. J Theor Biol 12 : 12–45

Harman D (1956) Aging: a theory based on free radical and radiation chemistry. J Gerontol Biol
    Sci 11 : 298–300

Harman D (1992) Free radical theory of aging. Mutation Res 275:257–266

Hass MA, Massaro D (1988) Regulation of the synthesis of superoxide dismutase in rat lungs during oxidant and hyperthermic stresses. J Biol Chem 263:776–781

Hayakawa M, Sugiyama S, Hattori K, Takasawa M, Ozawa T (1993) Age-associated damage in mitochondrial DNA in human hearts. Mol Cell Biochem 119:95–103

Hekimi S, Lakowski B, Barnes TM, Ewbank JJ (1998) Molecular genetics of life span in *C. elegans*: how much does it teach us? Trends Genet 14:14–20

Heydari AR, Wu B, Takahashi R, Strong R, Richardson A (1993) Expression of heat shock protein 70 is altered by age and diet at the level of transcription. Mol Cell Biol 13:2909–2918

Hodgkin J, Barnes TM (1991) More is not better: brood size and population growth in a self-fertilizing nematode. Proc R Soc Lond B Biol Sci 246:19–24

Hoffman AA, Parsons PA (1993) Selection for adult desiccation resistance in *Drosophila melanogaster*: Fitness components, larval resistance and stress correlations. Biol J Linn Soc 48:43–54

Hosokawa H, Ishii N, Ishida H, Ichimori K, Nakazawa H, Suzuki K (1994) Rapid accumulation of fluorescent material with aging in an oxygen-sensitive mutant *mev-1* of *Caenorhabditis elegans*. Mech Ageing Dev 74:161–170

Ishii N, Fujii M, Hartman PS, Tsuda M, Yasuda K, Senoo-Matsuda N, Yanase S, Ayusawa D, Suzuki K (1998) A mutation in succinate dehydrogenase cytochrome b causes oxidative stress and ageing in nematodes. Nature 394:694–697

Ishii N, Takahashi K, Tomita S, Keino T, Honda S, Yoshino K, Suzuki K (1990) A methyl viologen-sensitive mutant of the nematode *Caenorhabditis elegans*. Mutat Res 237:165–171

Ishii N, Suzuki N, Hartman PS, Suzuki K (1994) The effects of temperature on the longevity of a radiation-sensitive mutant *rad-8* of the nematode *Caenorhabditis elegans*. J Gerontol Biol Sci. 49:B117–120

Johnson TE, Hartman PS (1988) Radiation effects on life span in *Caenorhabditis elegans*. J Gerontol Biol Sci 43:B137–41

Johnson TE, Lithgow GJ, Murakami S (1996) Hypothesis: interventions that increase the response to stress offer the potential for effective life prolongation and increased health. J Gerontol Biol Sci 51A:B392–B395

Johnson TE (1990) *Caenorhabditis elegans* offers the potential for molecular dissection of the aging process. In: Schneider EL, Rowe JW (eds) Handbook of the Biology of Aging. Academic Press, New York, pp 45–59

Jonassen T, Proft M, Randez-Gil F, Schultz JR, Marbois BN, Entian KD, Clarke CF (1998) Yeast Clk-1 homologue (Coq7/Cat5) is a mitochondrial protein in coenzyme Q synthesis. J Biol Chem 273:3351–3357

Jones D, Russnak RH, Kay RJ, Candido EP (1986) Structure, expression, evolution of a heat shock gene locus in *Caenorhabditis elegans* that is flanked by repetitive elements. J Biol Chem 261:12006–12015

Kapahi P, Boulton ME, Kirkwood TB (1999) Positive correlation between mammalian life span and cellular resistance to stress. Free Radic Biol Med 26:495–500

Kennedy BK, Austriaco Jr NR, Zhang J, Guarante L (1995) Mutation in the silencing gene *SIR4* can delay aging in *S. cerevisiae*. Cell 80:485–496

Kenyon C, Chang J, Gensch E, Rudner A, Tabtiang R (1993) A *C. elegans* mutant that lives twice as long as wild type [see comments]. Nature 366:461–464

Khazaeli AA, Tatar M, Pletcher SD, Curtsinger JW (1997) Heat-induced longevity extension in Drosophila. I. Heat treatment, mortality, thermotolerance. J Gerontol A Biol Sci Med Sci 52:B48–B52

Kimura KD, Tissenbaum HA, Liu Y, Ruvkun G (1997) *daf-2*, an insulin receptor-like gene that regulates longevity and diapause in *Caenorhabditis elegans*. Science 277:942–946

Klass M, Hirsh D (1976) Non-ageing developmental variant of *Caenorhabditis elegans*. Nature 260:523–525

Klass MR (1977) Aging in the nematode *Caenorhabditis elegans*: major biological and environmental factors influencing life span. Mech Ageing Dev 6:413–429

Lakowski B, Hekimi S (1998) The genetics of caloric restriction in *Caenorhabditis elegans*. Proc Natl Acad Sci USA 95:13091–13096

Landry J, Chretien P, Lambert H, Hickel E, Weber LA (1989) Heat shock resistance conferred by expression of the HSP27 gene in rodent cells. J Cell Biol 109:7–15

Larsen PL (1993) Aging and resistance to oxidative damage in *Caenorhabditis elegans*. Proc Natl Acad Sci USA 90:8905–8909

Larsen PL, Albert PS, Riddle DL (1995) Genes that regulate development and longevity in *Caenorhabditis elegans*. Genetics 139:1567–1583

Lezza AM, Boffoli D, Scacco S, Cantatore P, Gadaleta MN (1994) Correlation between mitochondrial DNA 4977-bp deletion and respiratory chain enzyme activities in aging human skeletal muscles. Biochem Biophys Res Commun 205:772–779

Lin K, Dorman JB, Rodan A, Kenyon C (1997) *daf-16*: An HNF-3/forkhead family member that can function to double the life-span of *Caenorhabditis elegans*. Science 278:1319–1322

Lin YJ, Seroude L, Benzer S (1998) Extended life-span and stress resistance in the *Drosophila* mutant methuselah. Science 282:943–946

Linnane AW, Baumer A, Maxwell RJ, Preston H, Zhang CF, Marzuki S (1990) Mitochondrial gene mutation: the ageing process and degenerative diseases. Biochem Int 22:1067–1076

Lithgow GJ (1996) Invertebrate gerontology: the age mutations of *Caenorhabditis elegans*. Bioessays 18:809–815

Lithgow GJ, White TM, Hinerfeld DA, Johnson TE (1994) Thermotolerance of a long-lived mutant of *Caenorhabditis elegans*. J Gerontol Biol Sci 49:B270–B276

Lithgow GJ, White TM, Melov S, Johnson TE (1995) Thermotolerance and extended life span conferred by single-gene mutations and induced by thermal stress. Proc Natl Acad Sci USA 92:7540–7544

Liu Y, Guyton KZ, Gorospe M, Xu Q, Kokkonen GC, Mock YD, Roth GS, Holbrook NJ (1996) Age-related decline in mitogen-activated protein kinase activity in epidermal growth factor-stimulated rat hepatocytes. J Biol Chem 271:3604–3607

Maccubbin AE, Przybyszewski J, Evans MS, Budzinski EE, Patrzyc HB, Kulesz-Martin M, Box HC (1995) DNA damage in UVB-irradiated keratinocytes. Carcinogenesis 16:1659–1660

Malone EA, Inoue T, Thomas JH (1996) Genetic analysis of the roles of *daf-28* and *age-1* in regulating *Caenorhabditis elegans* dauer formation. Genetics 143:1193–1205

Martin GM, Austad SN, Johnson TE (1996) Genetic analysis of ageing: role of oxidative damage and environmental stresses. Nat Genet 13:25–34

Maynard Smith J (1958b) Prolongation of the life of *Drosophila subobscura* by brief exposure of adults to a high temperature. Nature 181:496–497

Maynard Smith J (1958a) The effects of temperature and of egg-laying on the longevity of *Drosophila subobscura*. J Exp Biol 35:832–843

Medawar PB (1952) An unsolved problem of biology. H.K. Lewis, London

Melov S, Lithgow GJ, Fischer DR, Tedesco PM, Johnson TE (1995) Increased frequency of deletions in the mitochondrial genome with age of *Caenorhabditis elegans*. Nucleic Acids Research 23:1419–1425

Melov S, Shoffner JM, Kaufman A, Wallace DC (1995) Marked increase in the number and variety of mitochondrial DNA rearrangements in aging human skeletal muscle. Nucleic Acids Res 23:4122–4126

Melov S, Hertz GZ, Stormo GD, Johnson TE (1994) Detection of deletions in the mitochondrial genome of *Caenorhabditis elegans*. Nucleic Acids Research 22:1075–1078

Miquel J (1992) An update on the mitochondrial-DNA mutation hypothesis of cell aging. Mutat Res 275:209–216

Miquel J, Economos J, Fleming J, Johnson Jr JE (1980) Mitochondrial role in cell aging. Experimental Gerontology 15:575–591

Morimoto RI, Tissieres A, Georgopoulos C (1994) The biology of heat shock proteins and molecular chaperones. Cold Spring Harbor Laboratory Press, New York

Morris JZ, Tissenbaum HA, Ruvkun G (1996) A phosphatidylinositol-3-OH kinase family member regulating longevity and diapause in *Caenorhabditis elegans*. Nature 382:536–539

Murakami S, Johnson TE (1996) A genetic pathway conferring life extension and resistance to UV stress in *Caenorhabditis elegans*. Genetics 143:1207–1218

Murakami S, Johnson TE (1998) Life extension and stress resistance in *Caenorhabditis elegans* modulated by the tkr-1 gene. Curr Biol 8:1091–1094

Ogg S, Paradis S, Gottlieb S, Patterson GI, Lee L, Tissenbaum HA, Ruvkun G (1997) The Fork head transcription factor DAF-16 transduces insulin-like metabolic and longevity signals in *C. elegans*. Nature 389:994–999

Orr WC, Sohal RS (1994) Extension of life-span by overexpression of superoxide dismutase and catalase in *Drosophila melanogaster*. Science 263:1128–1130

Paradis S, Ruvkun G (1998) *Caenorhabditis elegans* Akt/PKB transduces insulin receptor-like signals from AGE-1 PI3 kinase to the DAF-16 transcription factor. Genes Dev 12:2488–2498

Parkes TL, Elia AJ, Dickinson D, Hilliker AJ, Phillips JP, Boulianne GL (1998) Extension of Drosophila lifespan by overexpression of human SOD1 in motorneurons. Nat Genet 19:171–174

Parsell DA, Lindquist S (1994) Heat shock proteins and stress tolerance. In: Morimoto RI, Tissieres A, Georgopoulos C (eds) The biology of heat shock proteins and molecular chaperones. Cold Spring Harbor Laboratory Press, New York, pp 457–494

Parsons PA (1993) Evolutionary adaptation and stress: energy budgets and habitats preferred. Behav Genet 23:231–238

Parsons PA (1995) Inherited stress resistance and longevity: a stress theory of ageing. Heredity 75 (Pt 2):216–221

Parsons PA (1996) Rapid development and a long life: an association expected under a stress theory of aging. Experientia 52:643–646

Rollet E, Lavoie JN, Landry J, Tanguay RM (1992) Expression of *Drosophila's* 27 kDa heat shock protein into rodent cells confers thermal resistance. Biochem Biophys Res Commun 185:116–120

Rose MR, Vu LN, Park SU, Graves JLJ (1992) Selection on stress resistance increases longevity in *Drosophila melanogaster*. Exp Gerontol 27:241–250

Rouault JP, Kuwabara PE, Sinilnikova OM, Duret L, Thierry-Mieg D, Billaud M (1999) Regulation of dauer larva development in *Caenorhabditis elegans* by *daf-18*, a homologue of the tumour suppressor PTEN. Curr Biol 9:329–332

Sharma HK, Rothstein M (1980) Altered enolase in aged *Turbatrix aceti* results from conformational changes in the enzyme. Proc Natl Acad Sci USA 77:5865–5868

Sohal RS (1993) The free radical hypothesis of aging: An appraisal of the current status. Aging Clin Exp Res 5:3–17

Sun J, Tower J (1999) FLP recombinase-mediated induction of Cu/Zn-superoxide dismutase transgene expression can extend the life span of adult *Drosophila melanogaster* flies. Mol Cell Biol 19:216–228

Tan MW, Mahajan-Miklos S, Ausubel FM (1999) Killing of *Caenorhabditis elegans* by *Pseudomonas aeruginosa* used to model mammalian bacterial pathogenesis. Proc Natl Acad Sci USA 96:715–720

Tatar M, Khazaeli AA, Curtsinger JW (1997) Chaperoning extended life. Nature 390:30

Tissenbaum HA, Ruvkun G (1998) An insulin-like signaling pathway affects both longevity and reproduction in *Caenorhabditis elegans*. Genetics 148:703–717

Van Voorhies WA (1992) Production of sperm reduces nematode lifespan. Nature 360:456–458

Vanfleteren JR (1993) Oxidative stress and ageing in *Caenorhabditis elegans*. Biochem J 292:605–608

Vanfleteren JR, De Vreese A (1995) The gerontogenes *age-1* and *daf-2* determine metabolic rate potential in aging *Caenorhabditis elegans*. FASEB J 9:1355–1361

Varkey JP, Muhlrad PJ, Minniti AN, Do B, Ward S (1996) The *Caenorhabditis elegans spe-26* gene is necessary to form spermatids and encodes a protein similar to the actin-associated proteins kelch and scruin. Genes and Development 9:1074–1086

Walker GA, Walker DW, Lithgow GJ (1998) A relationship between thermotolerance and longevity in *Caenorhabditis elegans*. J Investig Dermatol Symp Proc 3:6–10

Walker GA, Walker DW, Lithgow GJ (1998) Genes that determine both thermotolerance and rate of aging in *Caenorhabditis elegans*. Ann NY Acad Sci 851:444–449

Williams GC (1957) Pleiotropy, natural selection, the evolution of senescence. Evolution 11:398–411

Yamashita N, Hoshida S, Nishida M, Igarashi J, Taniguchi N, Tada M, Kuzuya T, Hori M (1997) Heat shock-induced manganese superoxide dismutase enhances the tolerance of cardiac myocytes to hypoxia-reoxygenation injury. J Mol Cell Cardiol 29:1805–1813

Yamashita N, Hoshida S, Taniguchi N, Kuzuya T, Hori M (1998) Whole-body hyperthermia provides biphasic cardioprotection against ischemia/reperfusion injury in the rat. Circulation 98:1414–1421

Yasuda K, Adachi H, Fujiwara Y, Ishii N (1999) Protein carbonyl accumulation in aging dauer formation-defective (daf) mutants of *Caenorhabditis elegans*. J Gerontol A Biol Sci Med Sci 54:B47–B51

Yuh KC, Gafni A (1987) Reversal of age-related effects in rat muscle phosphoglycerate kinase. Proc Natl Acad Sci USA 84:7458–7462

Zeelon P, Gershon H, Gershon D (1973) Inactive enzyme molecules in aging organisms. Nematode fructose- 1, 6-diphosphate aldolase. Biochemistry 12:1743–1750

Oxidative Stress and Aging in *Caenorhabditis elegans*

Naoaki Ishii[1] and Philip S. Hartman[2]

1
Introduction

Aging is controlled by a complex interplay of both genetic and environmental factors. Because of this, many theories have been advanced that seek to explain the etiology of both cellular and organismal aging (Jazwinski 1996; Holliday 1997). In some cases, these theories are not mutually exclusive. One particularly popular theory posits that free radicals, especially those of molecular oxygen, can accelerate aging (Harman 1986). Martin et al. (1996) have eloquently articulated seven different classes of genetic loci that could modulate aging through oxidative damage. These include, for example, genes that would affect the generation of free radicals and others that modulate the scavenging of free radicals. A third class is those genes that specify repair enzymes. In addition, Cutler (1985) summarized a variety of data that support the notion that oxidative damage impacts aging. These include correlations between either free radical production or defenses against free radicals as compared with aging or life span. For example, the product of the standard metabolic rate and the maximum life span is roughly constant for various animals, indicating that animals with lower standard metabolic rates can live for longer periods than animals with higher rates. This concept, called the LEP (life-span energy potential), strongly suggests that oxygen radicals may exacerbate aging. More directly, the concentration of antioxidants in various mammalian tissues is inversely related to their maximum life-span potential (MLSP).

To reduce the damage induced by free radicals, organisms have developed multiple defense systems against these toxic species. The importance of these defense mechanisms is also well evidenced. For example, superoxide dismutase (SOD) activity is positively correlated in several organs with the maximum life span for various animal species, including primates. Similar

[1] Department of Molecular Life Science, Tokai University School of Medicine, Isehara, Kanagawa 259-1193, Japan
[2] Department of Biology, Texas Christian University, Fort Worth, TX 76129, USA

Results and Problems in Cell Differentiation, Vol. 29
Hekimi (Ed.): The Molecular Genetics of Aging
© Springer-Verlag Berlin Heidelberg 2000

correlations were observed for several other radical scavengers, including plasma urate, carotinoids, and vitamin E (Cutler 1985). In a particularly elegant experiment, Orr and Sohal (1994) showed that life span was significantly longer in transgenic flies carrying both SOD and catalase genes. In addition, while mice with no Sod1 or Sod3 were relatively normal, a defect in the mitochondrial enzyme Sod2 resulted in death at roughly 8 days (Carlson et al. 1995; Li et al. 1995; Reaume et al. 1996).

As is readily apparent, a genetic approach has been particularly instructive in illuminating the role played by oxygen free radicals in aging. A simple but profitable approach has been to isolate mutants with altered sensitivity to oxidative stress and then elucidate the molecular mechanisms that underpin these mutant phenotypes. The small, free-living nematode *Caenorhabditis elegans* has received much attention as a model to dissect the genetics (Riddle et al. 1997; Wood 1988), and by extension, the mechanisms by which oxidative stress triggers aging. *C. elegans* normally reproduces as a self-fertilizing hermaphrodite, which makes the isolation of mutants easy and allows for rapid inbreeding. Males also exist and can be used to construct stocks and position mutations on the genetic map. The sequence of the ca. 97-Mb genome has been almost completely determined (*C. elegans* Sequencing Consortium 1998). This nonpathogenic roundworm can grow in petri plates on a simple diet of *Escherichia coli*. A major advantage of using *C. elegans* for aging research is its mean life span of 10 to 20 days. Associated with this short life span is a rapid life cycle of approximately 3.5 days at 20°C.

Here we focus on the isolation and characterization of two short-lived mutants (*mev-1* and *rad-8*) that are hypersensitive to oxygen.

2
Genetics and Environmental Causes of Aging

On the basis of the life span variation among the progeny issuing from a cross between two inbred strains, Johnson and Wood (1982) estimated that the heritability of life span in *C. elegans* is between 20 and 50%. We took a somewhat different approach to address this issue. Using the completely inbred wild-type strain, N2, we determined the extent to which parental life span influenced the life span of their progeny. The data indicated that, despite the fact that there was some variation in life span from animal to animal, the parental life span did not dramatically influence progeny life span (Fig. 1). This provides further evidence that the life span is under genetic and environmental control; but what are these environmental factors? As described throughout this chapter, oxidative stress is most likely one of them.

Fig. 1a,b. Influence of parental life span on life span of progeny. Neither the average (a) or the maximum (b) life span of mothers had a strong influence on the life span of progeny, yet there was considerable variation in both average and maximum life spans. The *brackets* indicate two progeny derived from the same mother

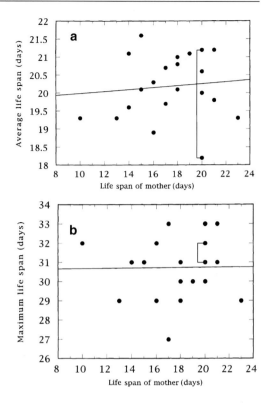

3
Isolation of Mutants

We isolated two methyl viologen (paraquat)-sensitive mutants, *mev-1* and *mev-2*, from ethyl methane sulfonate-mutagenized wild-type animals (Ishii et al. 1990). The toxic effects of this herbicidal drug on cells and animals are believed to be mediated by superoxide anions (Bagley et al. 1986). One paraquat-resistant mutant, *mev-3*, was also isolated by screening the mutator strain RW7000 (Yamamoto et al. 1996). In this strain, Tc1 and other transposable elements frequently insert into wild-type genes, creating spontaneous mutants. We also found that the radiation-sensitive mutant *rad-8* (Hartman and Herman 1982) is paraquat-sensitive (Ishii et al. 1993). In mutant isolation and testing, L1 larvae were cultured on plates containing various concentrations of paraquat. At the highest concentration (0.2 mM), most wild-type animals developed into L4 larvae or adults within 4 days. However, *mev-1* and *rad-8* usually arrested as L1 or L2 larvae. (Fig. 2a).

The *mev-1* and *rad-8* mutants were also oxygen-sensitive. On the other hand, *mev-2* was not oxygen-sensitive. Whereas wild-type animals developed

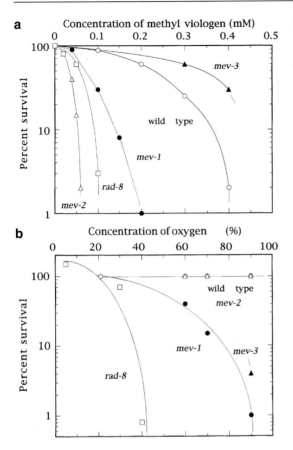

Fig. 2a,b. Sensitivities of wild type *mev-1*, *mev-2*, *mev-3*, and *rad-8* to methyl viologen (a) and oxygen (b)

nearly normally under 90% oxygen, these oxygen-sensitive mutants larvae arrested at L1 larval stage (Fig. 2b). Contrary to expectation, the paraquat-resistant *mev-3* mutant proved hypersensitive to oxygen. Thus, oxygen and paraquat sensitivities did not necessarily correlate.

   Each of these mutations behaved as single, recessive Mendelian factors. Three-factor crosses using visible genetic markers placed *mev-1* on chromosome III (Ishii et al. 1990, 1998); *mev-2* on chromosome X (unpubl. data); and *mev-3* and *rad-8* on chromosome I (Yamamoto et al. 1996; Hartman and Herman 1982).

4
Fecundity

Each of these oxygen-sensitive mutants had normal morphology under atmospheric oxygen. However, they had low vital activity; i.e., slow growth and low fecundity. For example, egg-laying in the wild type began on the 3rd day after hatching, reached a maximum of about 130 eggs per animal per day on the 5th day and continued for another week. The average number of eggs laid per individual was 287. Egg production was delayed for about half a day in the *mev-1* mutant and the average number of eggs laid per animals was only 77 (Ishii et al. 1990). Similar delays occurred with *rad-8* (Ishii et al. 1993).

5
Life Span

5.1
mev-1

The mean and maximum life spans of both wild type and *mev-1* decreased as oxygen concentrations were varied between 1 and 60%, but at markedly different rates (Ishii et al. 1990). Specifically, the mean and maximum life spans of wild type under atmospheric conditions (21% oxygen) were 26 days and 33 days, respectively. Life spans were extended significantly under 1% oxygen (mean, 30 days; maximum, 41 days), while those under 60% oxygen were shortened considerably (mean, 23 days; maximum, 28 days). Life spans were relatively constant within the range of 2 to 40%. Conversely, the mean and maximum life spans of the *mev-1* mutant were more profoundly affected by oxygen. Specifically, this mutant had mean and maximum life spans only 15% shorter than wild type when maintained at 1% oxygen (mean, 26 days, maximum, 35 days). However, the life spans were ca. 65% shorter than wild type when *mev-1* animals were tested under 60% oxygen (mean, 8 days; maximum, 10 days) (Honda et al. 1993).

The Gompertz component is a parameter of aging rate and was smaller for wild type under 1% oxygen than concentrations of 2% or greater. Larger increases were observed with *mev-1* (Fig. 3), indicating that oxygen negatively impacted life span and aging rate to a much greater degree in the *mev-1* mutant than in wild type. Furthermore, early exposure to 1% did not extend the life span if *mev-1* animals were incubated subsequently at higher concentrations. This indicates that the deleterious effects of oxygen were not exclusively secondary manifestations of perturbations that occurred during development and maturation (Honda et al. 1993).

5.2
rad-8

Interestingly, the life span of *rad-8* depended on both temperature and oxygen concentration (Ishii et al. 1993, 1994). At temperatures greater than

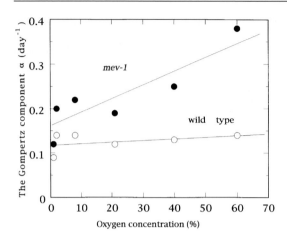

Fig. 3. The effect of oxygen concentration on the Gompertz components for wild type and *mev-1*

20°C and under atmospheric oxygen, the *rad-8* mutant possessed a life span roughly equivalent to that of the wild type. The life spans of this mutant varied substantially when oxygen concentrations were elevated. Surprisingly, at 16°C, the mutant actually lived significantly longer than did the wild type. This lengthened life span was due to slower development to maturity as opposed to a decreased rate of aging. It was also dependent on oxygen concentration, because the mean life spans of *rad-8* and wild type were experimentally identical when reared at 16°C in the presence of 5% oxygen rather than atmospheric oxygen. Thus, the *rad-8* mutant represents an interesting paradox, as its life span can either be shortened or lengthened relative to wild type, depending on the specific temperature and oxygen conditions.

6
Aging Markers

Fluorescent materials (lipofuscin) and protein carbonyl derivatives can be formed in vivo as a result of metal-catalyzed oxidation and accumulate during aging in disparate model systems (Strehler et al. 1959; Epstein et al. 1972; Stadman and Oliver 1991; Stadman 1992). As such, they have been employed as markers of aging, particularly as indicators of oxidation of lipids and proteins.

6.1
Fluorescent Materials

As observed using fluorescence microscopy, *C. elegans* contains blue auto-fluorescent granules and materials in its intestinal cells. These granules and materials differentially accumulated in wild type and *mev-1* in an age-dependent fashion (Hosokawa et al. 1994). Specifically, accumulation was identical over the first 5 days after hatching but was significantly higher in *mev-1* at later times. Fluorescent material in methanol/water extracts of both wild type and *mev-1* also accumulated with increasing age. As with auto-fluorescent materials, these fluorescent materials in *mev-1* accumulated more than in the wild type. For example, the amount was twice that of wild type when measured 10 days after hatching. Oxygen concentration affected accumulation rates. At low oxygen concentrations (2%) there was insignificant accumulation in either strain. Conversely, 90% oxygen had no influence on the accumulation rate in wild type but greatly accelerated the accumulation rate in the *mev-1* mutant (Fig. 4a).

6.2
Protein Carbonyls

The protein carbonyl contents in young wild-type and *mev-1* adults (4 to 8 days after hatching) were similar (Adachi et al. 1998). Afterwards, different

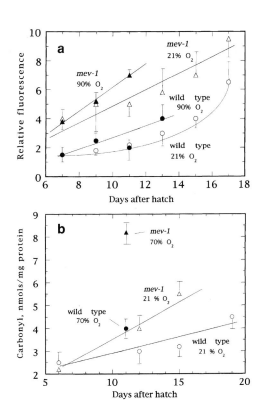

Fig. 4a,b. Accumulation of fluorescent compounds (a) and carbonyl compounds (b) in wild type and *mev-1* grown in either 21 or 90% oxygen

accumulations of carbonyl were observed with the genotype and increasing age. In wild type, carbonyl content accumulated in an age-dependent fashion and reached 4.7 nmol mg^{-1} protein at the time of death (20 days). Accumulation was significantly more rapid in *mev-1* and peaked at 5.7 nmol mg^{-1} protein at death (age 15–16 days). As with fluorescent materials, protein carbonyl levels were more impacted by elevated oxygen levels in *mev-1* than in wild type. For example, exposure to 70% oxygen between 4 and 11 days caused 100 and 31% increases in carbonyl in *mev-1* and wild type over the respective basal levels as compared with 21% oxygen, respectively (Fig. 4b).

It is important to note that these accumulation patterns mirror the effects of oxygen on life span, which supports the validity of using these as markers of aging.

7
Superoxide Dismutase (SOD) Activity

Dismutation of superoxide anion to hydrogen peroxide by SOD is a major antioxidant defense system. There are two isoforms of this enzyme, with one containing zinc and copper in its active center and the other containing manganese. The activities of these two types can be distinguished by adding potassium cyanide to the reaction mixture, which specifically inhibits the Zn/Cu enzyme. The SOD activity in *mev-1* was about 50% that of the wild type (Ishii et al. 1990). Most wild-type activity was inhibited by potassium cyanide, indicating that most SOD in wild type was the Zn/Cu isoform. In contrast, 14–21% of the residual activity in *mev-1* could be attributed to the Mn-SOD isozyme. As a caveat, these measurements were made from asynchronous populations and are conceivably affected by unrecognized differences in the age distributions of the two strains.

8
Molecular Cloning of *mev-1*

To clone *mev-1* (Ishii et al. 1998), the gene was first mapped between *unc-50* and *unc-49* on linkage group III. Cosmids from this region were then tested for their abilities to rescue *mev-1* mutants from oxygen hypersensitivity after germline transformation. Only cosmid T07C4 was able to rescue the *mev-1* mutant phenotype. By testing various subclones from this cosmid, a 5.6-kb fragment was identified that also restored resistance to high oxygen concentrations. This fragment included a putative gene, named *cyt-1*, that is homologous to bovine succinate dehydrogenase (SDH) cytochrome b$_{560}$

(GenBank accession number L26545). The *cyt-1* gene was sequenced from the *mev-1* strain and was found to contain a missense mutation resulting in change from glycine to glutamic acid.

9
Enzyme Activity of Cytochrome b$_{560}$

Electron transport is mediated by five multimeric complexes (complexes I–V) that are embedded in the inner membrane of the mitochondrion. Mitochondrial succinate-ubiquinone reductase (complex II), which catalyzes electron transport from succinate to ubiquinone, is composed of succinate dehydrogenase (SDH)(flavin protein: Fp and iron-sulfur protein: Ip) and two other subunits containing cytochrome b$_{560}$. In vivo, SDH is anchored to the inner membrane with cytochrome b$_{560}$ and is the catalytic component of complex II. Using separate assays, it is possible to quantify specifically both SDH activity and complex II activity. This was done with wild type and *mev-1* after extracts of each were subjected to differential centrifugation to separate mitochondria and mitochondrial membranes from cytosol (Ishii et al. 1998). The SDH activity in the *mev-1* mitochondrial fraction was experimentally identical to that of wild type. Conversely, complex II activity in the *mev-1* membrane fraction was reduced over 80%. As expected of a mitochondrial enzyme, no SDH activity was observed in the cytosol. Thus, the *mev-1* mutation affected neither SDH anchoring to the membrane nor SDH activity per se. However, it dramatically compromised the ability of complex II to participate in electron transport.

10
Mutagenesis

Oxygen-induced mutation frequencies in *C. elegans* were determined using the *fem-3* gene as a target (P. S. Hartman, B. Ponder and N. Ishii, unpubl.). As developed by Hartman and colleagues (1995), this system allows detection of rare loss-of-function mutations via their ability to relieve the temperature-sensitive sterility imposed by a gain-of-function mutation in *fem-3*. Frequencies were measured for wild type as well as for *rad-8* and *mev-1*. As expected, increases in oxygen (up to 90%) resulted in increased recovery of mutants in the wild-type background. Mutation frequencies were significantly elevated (ca. 5–10×) by the *mev-1* mutation at all oxygen concentrations tested (from 10–50%). Conversely, the mutation frequencies were unaffected by inclusion of the *rad-8* mutation.

There are several reasons why a mutation in *mev-1* might confer hypermutability to oxygen. First, as described above, *mev-1* encodes a subunit of the mitochondrial enzyme succinate dehydrogenase. This could result in excess superoxide anion production in the mitochondria which could have easily translated into more oxidative damage to the nucleus. Second, SOD levels are reduced in *mev-1* animals, making them less able to eliminate superoxide anion. Third, the *mev-1* mutation could result in reduced ATP production, which could then impact DNA repair fidelity. The presence of 8-oxoguanine has been measured in wild-type, *rad-8*, and *mev-1* animals (N. Ishii and H. Kasai, unpubl.). This DNA damage is generated at high frequencies by oxidative damage and is considered to be its primary mutagenic lesion (Grollman and Moriya 1993). Levels of 8-oxoguanine levels in the DNA were experimentally identical in the three strains when reared under atmospheric oxygen. It is possible that significantly more oxidative damage was induced in *mev-1* animals but was not reflected in steady-state levels, which were measured. For example, senescent human fibroblasts excised almost four times more 8-oxoguanine than did young cells yet accumulated only 35% more damage (Chen et al. 1995).

11
Apoptosis in *mev-1* and *rad-8* Mutants

Much effort has been directed at understanding the process and consequences of apoptosis (sometimes referred to as programmed cell death). In *C. elegans* a large number of specific cells undergo apoptosis during normal development (cf. Miller and Marx 1998 and associated articles). We have made three types of observation suggesting that, in addition to this normal and scheduled developmental phenomenon, oxygen can trigger massive apoptosis in *rad-8* and *mev-1* embryos.

First, significant numbers of *mev-1* and *rad-8* embryos contained many apoptotic nuclei when viewed under Nomarski DIC microscopy. The response was slightly different, depending upon the strain examined. Specifically, when exposed to 90% oxygen *rad-8* early embryos (<8 cells) typically arrested before the 16-cell stage and not produce apoptotic nuclei. Conversely, apoptosis was readily evident if *rad-8* embryos were exposed to high oxygen at the 50–300-cell stage. On the other hand, apoptosis in *mev-1* embryos was most prevalent if early embryos were exposed immediately to high oxygen. Apoptotic nuclei were not widespread in either mutant until 24 h after oxygen presentation. The number of apoptotic nuclei per embryo varied and approached 100% with extended incubation. Such embryos were indistinguishable from *ced-9(lf)* embryos, known to undergo widespread apoptosis.

Second, after exposure to high oxygen concentrations, DAPI-stained *mev-1* and *rad-8* embryos contained nuclei that had a punctuate appearance. The nuclei of these embryos also appeared indistinguishable from DAPI-stained *ced-9(lf)* embryos, which again served as a positive control. The kinetics correlated with Nomarski observations mentioned above.

Third, DNA extracted from *mev-1* and *rad-8* embryos yielded a nucleosomal ladder upon gel electrophoresis. As with the other two criteria, the ladder was visible only after 24–48 h of embryo incubation in high oxygen.

12
Mechanism of Cell Damage
by the *mev-1* Mitochondrial Abnormality

How, then, does the *mev-1* mutation exert its effects on mitochondria and, ultimately, the nematode? Cytochrome b_{560} is predicted to have three membrane-spanning domains. The substitution of glutamic acid for glycine is at position 71, only two amino acids removed from a histidine residue (His-73) that is thought to serve as a heme ligand. This could affect the ability of iron to accept and relinquish electrons, thus explaining the complex II deficiency in the *mev-1* mutant. As a consequence, the precocious aging and free-radical hypersensitivity of *mev-1* could result from two distinct mechanisms (Fig. 5). First, the mutation could cause electron transport to be deregulated such that oxygen uptake into mitochondria is higher than in wild type. This, in turn, could result in increased production of free radicals, primarily superoxide anion. In complex I (NADH-ubiquinone reductase) and complex III (ubiquinone-cytochrome c reductase), ubiquinone is considered as the main source of superoxide anion (Turrens and Boveris 1981; Turrens et al. 1985;

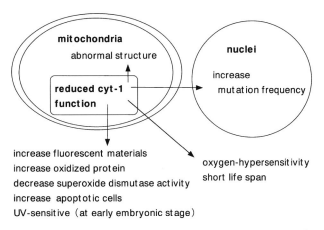

Fig. 5. Schematic representation of complex II, situated in the mitochondrial membrane. The *mev-1* mutation could confer hypersenstivity to oxidatative stress through either decreased ATP production or increased free radical generation

Sugioka et al. 1988). Especially, in coenzyme Q cycle, ubisemiquinone can be spontaneously oxidized to quinone with concomitant production of super-oxide anion (Turrens et al. 1985). In complex III, the function of cytochrome b is to remove the semiquinone by effectively acting as a dismutase in the Q cycle. It is possible that similar events occur in complex II. The *mev-1* mutation would then compromise the ability of Q semiquinone to dismutate, resulting in elevated production of superoxide anions.

Second, mutational perturbation of respiration could compromise ATP production and, as a result, lead to precocious aging (Fig. 5). Given that complex II plays a role in not only the electron transport but also the Krebs cycle, a deficiency in complex II may affect the energy metabolism by re-ducing ATP production. It has been postulated that such reductions could have major impacts on the fidelity of cellular defenses and repair processes (Shigenaga et al. 1994). Indeed, Lieberthal and colleagues (1998) showed that cells subjected to ATP depletion below 15% of the controls died uniformly of necrosis. Moreover, cells subjected to ATP depletion of between 25 and 70% of the controls died by apoptosis. Milder debts may lead to human myopathies and neurological diseases that include the presentation of precocious aging (Bourgeron et al. 1995). In addition, there is ample evidence that mitochondria contribute to apoptosis signaling via the production of reactive oxygen species (Mignotte and Vayssiere 1998). Blockage of energy metabolism can induce cell death, and Bcl-2 has been shown to efficiently inhibit the apoptosis induced by a metabolic inhibitor (Marton et al. 1997). Leakage of cytochrome c may mediate this response by leaking into the cytosol and activating the caspases that are essential for cell death (Kluck et al. 1997; Yang et al. 1997). The *mev-1* mutation is in cytochrome b_{560}, the *cyt-1* gene (Ishii et al. 1998). Interestingly, *ced-9*, a negative regulator of apoptosis in *C. elegans* and a homologue of Bcl-2, shares a common pro-motor with *cyt-1*. As discussed in Section 10, *mev-1* mutants are prone to massive apoptosis. It is also possible that ATP depression may lead cells to lactic acidosis. In turn, this may make cells more susceptible to oxidative stress. Thus, these observations serve to strengthen the evidence linking free radical damage and mitochondrial dysfunction to such phenomena as ap-optosis and aging (Fig. 6a,b).

13
Other *C. elegans* Life-Span Mutants Show Abnormal Responses to Oxidative Stress

In addition to *mev-1* and *rad-8*, other *C. elegans* mutants have also been isolated that possess altered life spans. Most of these show some abnormal-ities in their responses to oxidative stress. For example, *age-1* lives longer

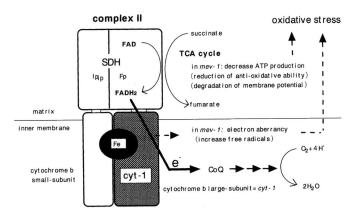

Fig. 6. Summary of the pleiotrophic phenotypes caused by the *cyt-1(mev-1)* mutation

than wild type and has proven to be oxidative stress-resistant (Friedman and Johnson 1988a,b; Larsen 1993; Murakami and Johnson 1996). This may be the result of age-associated increases in SOD and catalase activities (Larsen 1993; Vanfleteren 1993). Other long life-span mutants, *daf-2* (Kenyon et al. 1993), *spe-26* (Van Voorthies 1992; Gems and Riddle 1996) and *clk-1* (Wong et al. 1995) are also resistant to oxidative damage or other environmental stresses (Martin et al. 1996). Accumulation of protein carbonyl in long-lived *age-1* and *daf-2*, short-lived *mev-1*, and wild type are inversely correlated to the order of their comparative life spans (Adachi et al. 1998; Yasuda et al. 1999).

Recently, these genes have been cloned. The *age-1* gene is a subunit of a homologue of mammalian phosphatidylinositol-3-OH kinase (Morris et al. 1996). The *daf-2* gene encodes an insulin-like receptor (Kimura et al. 1997). These two genes play pivotal roles in signal transduction, specifically determining whether animals carry forward normal development or alternatively become dauer larvae, a "nonaging" developmental alternative. In addition, a third gene, *clk-1*, encodes the *C. elegans* homologue to the mitochondrial protein in coenzyme Q (CoQ) synthetase (Ewbank et al. 1997; Jonassen et al. 1998). Thus, a unifying theme of the genetics of aging in *C. elegans* is that mutations that both shorten or lengthen life span involve resistance or sensitivity to oxidative stress.

14
Closing Comments

Two final points are in order. First, as with *C. elegans*, human aging disorders also show pleiotropy. For example, Werner's syndrome presents some of the same manifestations as observed in *rad-8* and *mev-1* (Martin 1978). Second, since the pathology of precocious aging in *mev-1* and *rad-8* mimic that observed in wild type save for rate, it is reasonable to assume that a common (or at least related) mechanism(s) account for both. Therefore, further studies with these mutants should continue to illuminate the relationship between free radical damage and aging.

Acknowledgments. This work was supported in part by a Grant in Aid for Aging Research from the Ministry of Health and Welfare and by a Grant in Aid for Scientific Research form the Ministry of Education, Science, Sports and Culture, Japan.

References

Adachi H, Fujiwara Y, Ishii N (1998) Effects of oxygen on protein carbonyl and aging in *Caenorhabditis elegans* mutants with long (*age-1*) and short (*mev-1*) life spans. J Geront Biol Sci 53A:B240–B244

Bagley AC, Krall J, Lynch RE (1986) Superoxide meditates the toxicity of paraquat for Chinese hamster ovary cells. Proc Natl Acad Sci USA 93:3189–3193

Bourgeron T, Rustin P, Chretien D, Birch-Machin M, Bourgeois M, Viegas-Peqignot E, Munnich A, Rotig A (1995) Mutation of a nuclear succinate dehydrogenase gene results in mitochondrial respiratory chain deficiency. Nat Genet 11:144–149

Carlsson LM, Jonsson J, Edlund T, Marklund SL (1995) Mice lacking extracellular superoxide dismutase are more sensitive to hyperoxia. Proc Natl Acad Sci USA 92:6264–6268

C. elegans Sequencing Consortium (1998) Genome sequence of the nematode *C. elegans*: a platform for investigating biology. Science 282:2012–2018

Chen Q, Ficher A, Reagan JD, Yan L-J, Ames BN (1995) Oxidative DNA damage and senescence of human diploid fibroblast cells. Proc Natl Acad Sci USA 92:4337–4341

Cutler RG (1985) Antioxidants and longevity of mammalian species. In: Woodland AD, Blackett AD, Hollaender A (eds) Molecular biology of aging. Plenum Press, New York pp15–73

Epstein J, Himmelhoch S, Gershon D (1972) Studies on aging in nematodes. III. Electron microscopical studies on age-associated cellular damage. Mech Ageing Dev 1:245–255

Ewbank JJ, Barnes TM, Lakowski B, Lussier M, Bussey H, Hekimi S (1997) Structural and functional conservation of the *Caenorhabditis elegans* timing gene *clk-1*. Science 275:980–983

Friedman DB, Johnson TE (1988a) A mutation in the *age-1* gene in *Caenorhabditis elegans* lengthens life and reduces hermaphrodite fertility. Genetics 118:75–86

Friedman DB, Johnson TE (1988b) Three mutants that extended both mean and maximum lie span of the nematode, *Caenorhabditis elegans*, define the *age-1* gene. J Gerontol Biol Sci 43:B102–B109

Gems D, Riddle DR (1996) Longevity in *Caenorhabditis elegans* reduced by mating but not gamete production. Nature 379:723–725

Grollman AP, Moriya M (1993) Mutagenesis by 8-oxoguanine: an enemy within. Trends Genet 9:246–249

Harman D (1986) Free radical theory of aging: effect of free radical reaction inhibitors on the mortality rate of male LAF mice. J Gerontol 23:476–482

Hartman PS, Herman RK (1982) Radiation-sensitive mutants of *Caenorhabditis elegans*. Genetics 102:159–178

Hartman PS, DeWilde D, Dwarakanath V (1995) Genetic and molecular analyses of UV radiation-induced mutations in the *fem-3* gene of *Caenorhabditis elegans*. Photochem Photobiol 61:607–614

Hengartner MO, Horvitz HR (1994) *C. elegans* cell survival gene *ced-9* encodes a functional homologue of the mammalian proto-oncogene *bcl-2*. Cell 76:665–676

Holliday R (1997) Understanding aging. Philos Trans R Soc Lond B Biol Sci 352:1793–1797

Honda S, Ishii N, Suzuki K, Matsuo M (1993) Oxygen-dependent perturbation of life span and aging rate in the nematode. J Gerontol Biol Sci 48:B57–B61

Hosokawa H, Ishii N, Ishida H, Ichimori K, Nakazawa H, Suzuki K (1994) Rapid accumulation of fluorescent material with aging in an oxygen-sensitive mutant *mev-1* of *Caenorhabditis elegans*. Mech Ageing Dev 74:161–170

Ishii N, Takahashi K, Tomita S, Keino T, Honda S, Yoshino K, Suzuki K (1990) A methyl viologen-sensitive mutant of the nematode *Caenorhabditis elegans*. Mutat Res 237:165–171

Ishii N, Suzuki N, Hartman PS, Suzuki K (1993) The radiation-sensitive mutant *rad-8* of *Caenorhabditis elegans* is hypersensitive to the effects of oxygen on aging and development. Mech Ageing Dev 68:1–10

Ishii N, Suzuki N, Hartman PS, Suzuki K (1994) The effects of temperature-sensitive mutant *rad-8* of the nematode *Caenorhabditis elegans*. J Gerontol Biol Sci 49:B117–B120

Ishii N, Fujii M, Hartman PS, Tsuda M, Yasuda K, Senoo-Matsuda N, Yanase S, Ayusawa D, Suzuki K (1998) A mutation in succinate dehydrogenase cytochrome b causes oxidative stress and ageing in nematodes. Nature 394:694–697

Jazwinski SM (1996) Longevity, genes, and aging. Science 273:54–59

Johnson TE, Wood WB (1982) Genetic analysis of life span in *Caenorhabditis elegans*. Proc Natl Acad Sci USA 79:6603–6607

Jonassen T, Proft M, Randez-Gil F, Schultz JR, Marbois BN, Entian K-D, Clarke CF (1998) Yeast *clk-1* homologue (coq7/cat5) is a mitochondrial protein in coenzyme Q. J Biol Chem 273:3351–3357

Kenyon C, Chang J, Gensch E, Rudner A, Tabtang R (1993) A *C. elegans* mutant that lives twice as long as wild type. Nature 366:461–464

Kimura KD, Tissenbaum HA, Liu Y, Ruvkun G (1997) *daf-2*, an insulin receptor-like gene that regulates longevity and diapause in *Caenorhabditis elegans*. Science 277:942–946

Kluck CN, Bossy-Wetzel E, Green DR, Newmeyer DD (1997) The release of cytochrome c from mitochondria: a primary site for Bcl-2 regulation of apoptosis. Science 275:1132–1136

Larsen PL (1993) Aging and resistance to oxidative damage in *Caenorhabditis elegans*. Proc Natl Acad Sci USA 90:8905–8909

Li Y, Huang TT, Carlson EJ, Melov S, Ursell PC, Olson JL, Noble LJ, Yoshimura MP, Berger C, Chan PH, Wallace DC, Epstein CJ (1995) Dilated cardiomyopathy and neonatal lethality in mutant mice lacking manganese superoxide dismutase. Nat Genet 11:376–381

Lieberthal W, Menza SA, Levine JS (1998) Graded ATP depletion can cause necrosis or apoptosis of cultured mouse proximal tubular cells. Am J Physiol 274:F315–F327

Martin GM (1978) Genetic syndromes in man with relevance to the pathology of aging. In: Bergsma D, Harrison DE (eds) Genetic effects on aging, birth defects. Original Article Series vol 14, Alan R Liss, New York, pp5–39

Martin GM (1991) Genetic and environmental modulations of chromosomal stability: their roles in aging and oncogenesis. Ann NY Acad Sci 621:401–417

Martin GM, Austad SN, Johnson TE (1996) Genetic analysis of ageing: role of oxidative damages and environmental stresses: review. Nat Genet 13:25–33

Marton A, Mihalik R, Bratincsak A, Adleff V, Petak I, Vegh M, Bauer PI, Krajcsi P (1997) Apoptotic cell death induced by inhibitors of energy conservation Bcl-2 inhibits apoptosis downstream of a fall of ATP level. Eur J Biochem 250:467–475

Mignotte B, Vayssiere J-L (1998) Mitochondria and apoptosis: review. Eur J Biochem 252:1–15

Morris JZ, Tissenbaum HA, Ruvkun G (1996) A phosphatidylinositol-3-OH kinase family member regulating longevity and diapause in *Caenorhabditis elegans*. Nature 382:536–539

Miller LJ, Marx J (1998) Apoptosis. Science 281:1301–1326

Murakami S, Johnson TE (1996) A genetic pathway conferring life extension and resistance to UV stress in *Caenorhabditis elegans*. Genetics 143:1207–1218

Orr WC, Sohal RS (1994) Extension of life span by over expression of superoxide dismutase and catalase in *Drosophila melanogaster*. Science 263:1128–1130

Reaume AG, Elliott J, Hoffman EK, Kowall NW, Ferrante RJ, Siwek-DF, Wilcox HM, Flood DG, Beal MF, Brown RH Jr, Scott RW, Snider WD (1996) Motor neurons in Cu/Zn superoxide dismutase-deficient mice develop normally but exhibit enhanced cell death after axonal injury. Nat Genet 13:43–47

Riddle DL, Blumenthal T, Mayer BJ, Priess JR (eds) (1997) *C. elegans* II. Cold Spring Harbor Laboratory Press, Cold Spring Harbor, New York

Shigenaga MK, Hagen TM, Ames BN (1994) Oxidative damage and mitochondrial decay in aging. Proc Natl Acad Sci USA 91:10771–10778

Stadman ER (1992) Protein oxidation and aging. Science 257:1220–1224

Stadman ER, Oliver CN (1991) Metal-catalyzed oxidation of proteins. J Biol Chem 266:2005–2008

Strehler BL, Mark DD, Mildvan AS, Gee MV (1959) Rate and magnitude of age pigment accumulation in the human myocardium. J Gerontol 14:257–264

Sugioka K, Nakano M, Totsune-Nakano H, Minakami H, Tero-Kubota S, Ikegami Y (1988) Mechanism of O_2-generation in reduction and oxidation cycle of ubiquinones in a model of mitochondrial electron transport systems. Biochem Biophys Acta 936:377–385

Turrens JF, Boveris A (1981) Generation of superoxide anion by the NADH dehydrogenase of bovine heart mitochondria. Biochem J 191:421–427

Turrens JF, Alexandre A, Lehninger AL (1985) Ubisemiquinone is the electron donor for superoxide formation by complex III of heart mitochondria. Arch Biochem Biophys 237:408–414

Vanfleteren JR (1993) Oxidative stress and ageing in *Caenorhabditis elegans*. Biochem J 292:605–608

Van Voorthies WA (1992) A production of sperm reduces nematode lifespan. Nature 360:456–458

Wong A, Boutis PA, Hekimi S (1995) Mutations in the *clk-1* gene of *Caenorhabditis elegans* affect development and behavioral timing. Genetics 139:1247–1259

Wood WB (ed) (1988) The nematode *Caenorhabditis elegans*. Cold Spring Harbor Laboratory Press, Cold Spring Harbor, New York

Yamamoto K, Honda S, Ishii N (1996) Properties of an oxygen-sensitive mutant *mev-3* of the nematode *Caenorhabditis elegans*. Mutat Res 358:1–6

Yang J, Liu X, Bhalla K, Kim CN, Ibrado AM, Cai J, Peng TI, Jones DP, Wang X (1997) Prevention of apoptosis by Bcl-2: release of cytochrome c from mitochondrial block. Science 257:1129–1132

Yasuda K, Adachi F, Fujiwara Y, Ishii N (1999) Protein carbonyl accumulation in aging dauer formation-defective (*daf*) mutants of *Caenorhabditis elegans*. J Gerontol Biol Sci 54A:B47–B51

Mutation Accumulation In Vivo and the Importance of Genome Stability in Aging and Cancer

Martijn E. T. Dollé, Heidi Giese, Harry van Steeg, and Jan Vijg*

1
Introduction

Somatic mutations are generally considered as the major cause of cancer. This can be derived from various observations, including the actual presence of mutations in the tumor genome in genes thought to be critically involved in tumor initiation and/or progression, such as *TP53*, *KRAS*, and *RB1* (Lengauer et al. 1998). Both exogenous (Greenblatt et al. 1994) and endogenous (Jackson et al. 1998) mutagenic mechanisms have been implicated in the induction of these mutations. Modeling of such gene defects into the mouse genome often results in accelerated tumorigenesis (Vijg and van Steeg 1998), confirming the critical role of specific mutations as a cause of cancer. Other indirect evidence involves the general observation that most, if not all, mutagens are also carcinogens and that heritable mutations in genes controlling genome stability pathways often confer a high cancer susceptibility (Vijg and van Steeg 1998).

Somatic mutations are thought to activate and/or inactivate specific genes involved in such processes as cell cycle regulation and growth control, tissue invasion, neovascularization, and DNA damage processing itself (i.e., to acquire a mutator phenotype; Loeb 1991). The accumulation of the successive mutations that result in cancer must take place stepwise, with each mutation having a selective advantage (Fearon and Vogelstein 1990). However, thus far, it is unclear how many different mutations are required for each particular tumor and the types of mutations (e.g., point mutations, genome rearrangements, aneuploidy) that are most important. A complicating factor in this respect is that most tumor genomes appear to contain numerous genomic alterations, of which only a part may be functionally important (Murakami and Sekiya 1998).

* Corresponding author.
University of Texas Health Science Center and CTRC Institute for Drug Development, 8122 Datapoint Drive, Suite 700, San Antonio, TX 78229, USA
Laboratory of Health Effects Research, National Institute of Public Health and the Environment, PO Box 1, 3720 BA Bilthoven, The Netherlands

Results and Problems in Cell Differentiation, Vol. 29
Hekimi (Ed.): The Molecular Genetics of Aging
© Springer-Verlag Berlin Heidelberg 2000

Somatic mutations have also been considered as a major cause of aging. Indeed, it has been pointed out that the exponential increase in cancer incidence with age is dependent not so much on chronological age as on biological age (Peto et al. 1975). While, for example, in the mouse, cancer is very frequent at 2–3 years of age, in humans the peak seems to occur no earlier than at 60–90 years. This difference is associated with an about 30-fold difference in life span and is likely to be due to more sophisticated genome stability systems in the longer-lived species (Hart et al. 1979). Since there is no evidence for a much higher mutation frequency in cells from young rodents as compared to young humans, this striking difference has been explained by a more rapid mutation accumulation in the rodent somatic DNA. Indeed, somatic mutation might be a common mechanism underlying both aging and cancer (Failla 1958; Szilard 1959; Curtis 1963; Burnet 1974). As argued by Martin (1991), an important part of the spectrum of degenerative changes associated with aging, is a loss of proliferative homeostasis, which may result in tissue atrophy as well as hyperplasia and neoplasia.

Thus far there is no direct evidence for a role of somatic mutations in aging per se. Unlike with tumor tissue, potentially causal mutations have not been demonstrated in old tissue. Indeed, since aged tissue represents a mosaic of individual cells rather than the clonal outgrowth of one or few cells, a functional link between somatic mutations and aging has proved more difficult to establish than in the case of cancer. More recently, evidence has been obtained for a role of telomere shortening, i.e., loss of minisatellite repeat copies in the absence of the enzyme telomerase, in replicative senescence (Bodnar et al. 1998). (Replicative senescence is the gradual loss of proliferative potential of normal human cells in culture.) However, a functional relationship, if any, between replicative senescence and organismal aging is as yet unclear (Campisi 1996). Nevertheless, telomere maintenance could be important for organismal aging, as exemplified by the recent demonstration of shortened life span, hematopoietic ablation, reduced capacity to respond to stresses, as well as increased incidence of spontaneous malignancies of telomerase null mice (Rudolph et al. 1999). These demonstrated effects of telomere instability and the accompanying genetic instability underscore that loss of genome integrity in vivo can lead to the loss of proliferative homeostasis as predicted by Martin (1991). However, it does not tell us much about mutational effects on aging in postmitotic cells. To test the hypothesis that mutation accumulation in vivo is the ultimate cause of cancer and other degenerative aspects of the aging process, model systems are needed that permit the quantitation and characterization of genomic mutations in different organs and tissues at different backgrounds of genome stability.

2
In Vivo Model Systems for Measuring Mutations

Cytogenetic analysis and the use of selectable marker genes, such as HPRT and HLA, have been very useful in establishing a pattern of age-related mutation accumulation in, mainly, human and mouse lymphocytes (Ramsey et al. 1995; Albertini and Hayes 1997). However, such methods do not permit the accurate quantitation and characterization of mutation in different organs and tissues over the lifetime of an animal. With the advent of transgenic animal technology, a completely different approach has become possible. Now a decade ago we reported the generation of the first transgenic mouse lines harboring *lacZ* reporter genes integrated in one or more chromosomes as part of a bacteriophage lambda vector that could be recovered and inspected in *E. coli* for mutations (Gossen et al. 1989). Initially, mutant *lacZ* genes were detected as colorless plaques among the wild-type blue plaques. Later, we developed a positive selection system based on an *E. coli* host with an inactivated *galE* gene (Gossen and Vijg 1993a). Upon receiving a wild-type *lacZ* gene such cells, grown in the presence of the lactose analogue phenyl-β-D-galactoside (p-Gal), produce UDP-galactose which is highly toxic when it cannot be converted into UDP glucose. Thus, on medium containing p-Gal only those cells infected with a mutant *lacZ* gene can give rise to a plaque. This greatly decreased the cost and time involved in the assay. An alternative model based on the same principle, but with the *lacI* gene as the reporter, was generated by Short and co-workers (Kohler et al. 1991). In this model mutational inactivation of the gene encoding the *lacI* repressor leads to derepression of the *lacZ* gene, resulting in β-gal expression and a blue plaque among colorless wild-type plaques.

Both the *lacZ* and *lacI* bacteriophage lambda system have now been extensively used for studying mutations in vivo, mainly for genetic toxicological purposes. Comparisons have been made in one and the same animal between the transgene locus and the HPRT (Skopek et al. 1995) or *Dlb-1* (Tao et al. 1993) locus. (The *Dlb-1* locus assay detects the loss of the binding site for the lectin *Dolichos biflorus* on the cell surface of the epithelial cells of the small intestine in *Dlb-1^a/Dlb-1^b* heterozygous mice.) It was found that transgene and natural gene reacted in essentially the same way to treatment with a mutagen. However, in blood lymphocytes a much lower background value was reported for the HPRT as compared to the (*lacI*) transgene (Skopek et al. 1995). Explanations for this could be the loss of HPRT mutations due to selection in vivo or in vitro against HPRT mutant cells, or the much higher methylation level of the transgene (which is not expressed). Deamination of 5′-methylcytosine is likely to be a major mechanism of spontaneous mutagenesis at such sites (Gossen and Vijg 1993b).

Another difference between the reporter transgene and an endogenous gene (in this case *Dlb-1*; Tao et al. 1993) is the much lower sensitivity of the former for X-irradiation. In part, this can be explained by the fact that large deletion mutations will go undetected as a consequence of the minimum vector size required for efficient packaging of bacteriophage lambda vectors (i.e., between 42 and 52 kb). In addition, genome rearrangements involving the mouse flanking regions and therefore one of the cos sites cannot be detected. DNA rearrangements leading to size-change mutants involving more than a few basepairs are likely to mainly involve illegitimate recombinations, which may have a preference for the mouse-flanking copies of the integrated concatemers. However, such deletion events are expected to occur, both spontaneously and induced by clastogens, as they have been observed at the HPRT locus and other selectable genes. Since mutations, induced in the genome at random, can be expected to have a higher functional impact when they involve large structural alterations than point mutations, it is especially for this reason that we decided to develop an alternative system based on integrated plasmid rather than bacteriophage lambda vectors.

3
The *lacZ*-Plasmid Mouse Model for Mutation Detection

As in the bacteriophage lambda model, also in the plasmid model the vector with the reporter gene *lacZ* is integrated in multiple copies in a head-to-tail organization (see Boerrigter et al. 1995; Fig. 1). The reason that initially plasmid vectors were not used as the vector of choice for transgenic mutation models is the notoriously low transformation efficiencies obtained with plasmids excised from their integrated state in the mammalian genome. Efficient intramolecular ligation of excised plasmids requires extensive dilution and the presence of the genomic DNA will also negatively influence the transformation efficiency. To solve this problem, methods were developed to recover the plasmids from genomic mouse DNA using magnetic beads coupled to the lac repressor protein, which selectively binds to the operator sequence in front of the *lacZ* gene (Gossen et al. 1993). After ligation, plasmids are introduced into *galE⁻ E. coli* host cells by means of electroporation. The p-Gal positive selection is identical to that for bacteriophage lambda described above; only mutant *lacZ* plasmids give rise to a colony. The rescue and mutant frequency determination procedure is schematically depicted in Fig. 1.

The characteristics of plasmid rescue should allow detection of a broad range of mutational events, including internal deletions as well as deletions with one breakpoint in the *lacZ* mutational target gene and the second breakpoint in the 3′ region flanking the concatemer. This is illustrated in

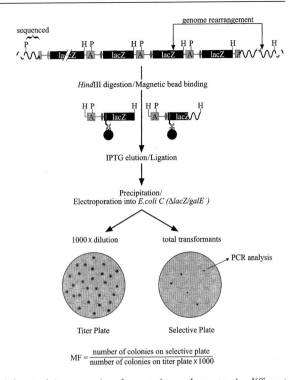

Fig. 1. Transgenic mouse model for studying mutation frequencies and spectra in different organs and tissues in relation to aging and cancer. Plasmids are integrated head to tail in the mouse genome. Several transgenic lines with the plasmid cluster at different chromosomal positions have been studied (Dollé et al. 1997). The plasmids are recovered from their integrated state by excision with a restriction enzyme, circularized by ligation, and subsequently used to transform *E. coli* cells. A very small part is plated on the titer plate, which contains X-gal, resulting in a blue halo round the colonies. The rest is plated on p-Gal plates to select for the mutants, which appear as red colonies because of the presence of tetrazolium. The figure also indicates how genome rearrangement events with one breakpoint in a *lacZ* gene and the other in the 3'-mouse flanking sequence can be recovered and detected. (For further details, see Boerrigter et al. 1995)

Fig. 1, showing excision of the plasmid from the mouse chromosomal DNA by using the restriction enzyme Hind III. Point mutations and deletion mutations in internal copies of the plasmid cluster are recovered on the p-Gal selective plates, as long as the origin of replication, the operator sequence, and the ampicillin resistance gene are not affected (Fig. 2A,B). Large deletions that involve the 3' mouse flanking region can also be recovered. In such cases a fragment will be recovered that includes a mouse sequence, from the breakpoint in the flanking region to the nearest Hind III restriction site. As a consequence, these recovered plasmids should hybridize with labeled mouse total genomic DNA, which appeared to be the case (Fig. 2C). Such mutants were called mouse-sequence mutants. Using this system we have analyzed

A

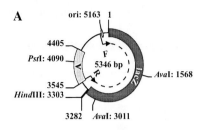

B a b c d e f g h i j k l m n o

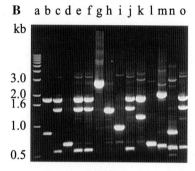

Fig. 2. A The plasmid pUR288 with the *lacZ* reporter gene, the ampicillin resistance gene, and the origin of replication. The HindIII, PstI, and AvaI restriction sites are indicated, as well as the forward (*F*) and reverse (*R*) primer for PCR amplification of the entire *lacZ* region. **B** Ethidium bromide-stained gel patterns of AvaI-digested PCR-amplified *lacZ* genes from a number of different mutant colonies taken from the selective plates. Point mutants show a wild-type banding pattern (three bands of 1.8, 1.4, and 0.57 bp). Any deviation from this pattern indicates a size-change mutant (lanes *b*, *d*, *g*, *h*, *i*, *k*, *l*, *m* and *n*). Lane *a* shows the 1-kb DNA size marker. **C** Hybridization of the pattern transferred to a nylon membrane with radiolabeled non-transgenic mouse genomic DNA. Mutants with a captured mouse sequence (lanes *g*, *h*, and *m*) represent genome rearrangements

C a b c d e f g h i j k l m n o

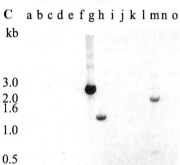

in vivo mutation accumulation after treatment with genotoxic agents and during the aging process (Boerrigter et al. 1995; Gossen et al. 1995; Dollé et al. 1997).

4
Monitoring Mutation Accumulation in Mice with Defects in Genome Stability Pathways

Mouse mutants carrying defined defects in genome stability systems are now being generated at an increasing pace. As expected, most of these knockout mice are predisposed to cancer. However, several knockout mice involve genes that in humans not only cause increased cancer predisposition, but also symptoms of accelerated aging, i.e., the segmental progeroid syndromes, such as Werner's syndrome (Yu et al. 1996), Cockayne's syndrome (van Gool et al. 1997), and ataxia telangiectasia (Shiloh and Rotman 1996). (The fact that heritable mutations in such genes cause accelerated aging supports the notion that genome stability systems are major longevity assurance systems.) In some of these diseases a high level of genomic instability has been demonstrated in peripheral blood lymphocytes of the patients, suggesting that an accelerated accumulation of somatic mutations might be the ultimate cause of the disease (Fukuchi et al. 1989). We have begun to cross-breed the plasmid *lacZ* mouse model for detecting mutations with DNA repair mouse mutants. Eventually, each of these hybrids will harbor a specific mutation in a genome stabilization pathway as well as the *lacZ* reporter gene for detecting somatic mutations.

4.1
The *TP53* Gene

A major category of genome stability systems involves cell cycle checkpoints, that is, monitoring systems for DNA damage that temporarily halt replication until the lesions are repaired (Hartwell and Kastan 1994). The first cell cycle checkpoint gene that is understood in some detail is the *TP53* gene. The *TP53* gene is the tumor suppressor gene most frequently mutated in human tumors. Its gene product has a central role in the G1 checkpoint and is thought to act, at least partly, through the induction of p21, an inhibitor of cyclin-dependent kinases and PCNA (proliferating cell nuclear antigen). Various signals of cellular distress, including DNA damage (most notably DNA double-strand breaks), hypoxia, and low levels of nucleotide triphosphate pools, use p53 as the trigger to a response (Levine 1997). A major component of this response clearly is cell cycle arrest. However, induction of p53 can also result in apoptosis, a form of programmed cell death (Lowe et al. 1993).

Crosses were made between the *lacZ*-plasmid mouse model and the *TP53* knockout mouse developed by Jacks et al. (1994). Subsequently, liver and spleen of the homozygous knockouts ($TP53^{-/-}$), as well as of the heterozygous animals ($TP53^{+/-}$), with the *lacZ*-plasmid vector also present, were analyzed for mutation frequencies. When sacrificed at early age, i.e., about 2 months, the C57Bl/6-based $TP53^{-/-}$ animals are normal. At a later age, i.e., between 2 and 7 months, they get cancer and die (Jacks et al. 1994). The $TP53^{-/-}$ animals were sacrificed at various time points to study the mutant frequency over the life span until about 7 months. As shown in Fig. 3, at early age mutant frequencies in liver and spleen were not different from the heterozygous *TP53* mutants or the control animals. Indeed, after treatment with 100 mg kg^{-1} bodyweight of the powerful mutagen ethyl nitrosourea equal amounts of

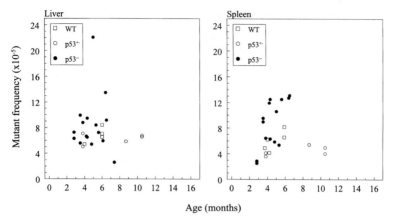

Fig. 3. Spontaneous mutant frequencies in liver and spleen of p53-deficient *lacZ* transgenic mice with age

mutations were induced in these young $TP53^{-/-}$ mice and the controls (Giese et al., submitted). However, at somewhat older age spontaneous mutant frequencies in the $TP53^{-/-}$ animals rapidly increases with a high individual variation (Fig. 3). This was statistically significant for the spleen, but not for the liver, and appeared to be a characteristic of the normal tissues, since histopathological analysis indicated no sign of tumors in the organ samples taken for mutant frequency determinations (Giese et al., submitted).

In some of the tumors in the $TP53^{-/-}$ animals, which were also analyzed (mainly from thymus and lymph node), somewhat higher mutant frequencies than those found in the normal tissues were observed (Fig. 4). Further characterization showed that in most cases these mutations involved point mutations. In this respect, it should be noted that tumor genomes from $TP53^{-/-}$ mice are cytogenetically highly unstable and characterized by excessive aneuploidy and other chromosomal aberrations (Smith and Fornace 1995). Thus, while $TP53$ is important in maintaining chromosomal stability, it is apparently much less important in maintaining stability at the nucleotide level.

Based on the results thus far, $TP53$ does not seem to play the major role in mutation induction as its function as "guardian of the genome" might have suggested. This is in spite of its important role in carcinogenesis. This seeming discrepancy can be explained by the assumption that $TP53$'s early role as a cell cycle checkpoint and in apoptosis of cells with severe DNA damage can be taken over by alternative pathways. At later ages, however, mutations are accumulating at an accelerated pace as compared to the normal aging process. This is accompanied by an increased predisposition to tumor formation. Apparently, at this adult stage the role of $TP53$ in apoptosis is important in preventing the accumulation of mutated and tumorigenic cells.

Fig. 4. Average spontaneous mutant frequencies in liver, spleen and tumor tissue of p53 deficient *lacZ* transgenic mice as compared to liver and spleen from *lacZ* control mice. Means for liver and spleen were calculated over all data points in Fig. 3; as a control group WT and $p53^{+/-}$ data were combined. *Error bars* indicate the standard deviation of the mean. *LN* lymph node; *Th* thymus. Significant differences were found between mutant frequencies in liver and spleen (either $p53^{-/-}$ or control) and tumor tissue from p53-deficient mice ($p < 0.038$, two-tail Mann-Whitney test)

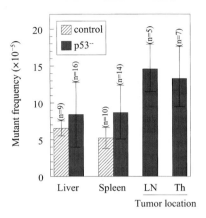

4.2
The *XPA* Nucleotide Excision Repair Gene

A major DNA repair mechanism in mammalian cells is nucleotide excision repair (NER). This system entails multiple steps that employ a number of proteins to eliminate a broad spectrum of structurally unrelated lesions such as UV-induced photoproducts, chemical adducts, as well as intrastrand cross-links and some forms of oxidative damage (Bootsma et al. 1998). Deficiency in NER has been shown to be associated with human inheritable disorders, such as xeroderma pigmentosum (XP), Cockayne's syndrome (CS), and trichothiodystrophy (TTD). These disorders are characterized by UV sensitivity, genomic instability, and various signs of premature aging (Bootsma et al. 1998). XP patients can be classified into at least seven complementation groups (*XPA-XPG*). One of the most frequently found groups in human is *XPA*. Its deficiency causes a complete block in NER and some severe symptoms, including a >2000-fold increased frequency of UVB-induced skin cancer and accelerated neurodegeneration (Hoeijmakers 1994; Cleaver and Kraemer 1995). The *XPA*-protein, in combination with replication protein A, has been proposed to be involved in DNA damage recognition, i.e., the pre-incision step of NER (Sugasawa et al. 1998).

Mice deficient in the NER gene *XPA* have been generated by gene targeting in embryonic stem cells (de Vries et al. 1995). These NER-deficient mice were demonstrated to mimic the phenotype of humans with xeroderma pigmentosum, that is, increased sensitivity to UVB and dimethylbenz[a]anthracene (DMBA)-induced skin cancer. This was found to be associated with an almost complete lack of UV-induced excision repair, measured by unscheduled DNA synthesis in cultured fibroblasts, and a much lower survival of fibroblasts after UV-irradiation or treatment with DMBA (de Vries et al. 1995). However, at early age NER-deficient mice do not show spontaneous abnormalities. These mice develop normally, indistinguishable from wild-type mice. The lack of spontaneous abnormalities in young mice might be due to the fact that under normal conditions mice have only limited exposure to NER-mediated DNA damage. Hence, at early age NER appears to be dispensable. This is in contrast to base excision repair, the inactivation of which is lethal (Gu et al. 1994).

In order to test the hypothesis that loss of NER causes accelerated mutation accumulation, preceding the onset of accelerated tumor formation and aging, XPA-deficient mice were crossed with the *lacZ* transgenic mice previously used to monitor mutation accumulation in liver and brain during aging (Dollé et al. 1997). In the hybrid $XPA^{-/-}$, *lacZ* mice, mutant frequencies were analyzed in liver and brain at early ages, 2 months, 4 months, and 9–12 months, and compared to the normal *lacZ* control background (Giese et al. 1999). As shown in Fig. 5, in liver of 2-month-old mice, mutant frequencies were at a normal level, as established earlier for liver DNA of animals in this age range (Dollé et al. 1997). In 4-month-old $XPA^{-/-}$, *lacZ* mice, mutant frequencies were significantly increased by a factor of 2. A further, albeit smaller increase was observed between 4 and 9–12 months. At the latter age level mutant frequencies were in the range of the maximum level reported earlier for 25–34-month-old *lacZ* control mice, with a similar high individual variation (Dollé et al. 1997). In brain, mutant frequencies were not found to increase over the age levels studied, which is in keeping with the lack of an age-related increase reported previously for the *lacZ* control mice (Dollé et al. 1997). Mutant spectra were analyzed from liver of 2-month and 9–12-month- old $XPA^{-/-}$, *lacZ* mice, indicating an about equal fraction of size change versus point mutations (Giese et al. 1999).

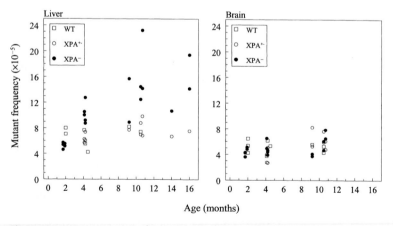

Fig. 5. Spontaneous mutant frequencies in liver and brain of XPA-deficient *lacZ* transgenic mice with age

The results of this study indicate that a deficiency in the NER gene *XPA* causes an accelerated, early accumulation of somatic mutations in liver, but not in brain. Since complete life-span studies of the *XPA*-deficient mice have not yet been completed, it is difficult to say whether or not the defect is associated with signs of premature aging. At least until the age of about 15 months, the animals do not appear to be much different from the control *XPA*-proficient mice. Interestingly, at older age, i.e., from about 15 months, onwards, $XPA^{-/-}$ mice show an increased frequency of hepatocellular adenomas (de Vries et al. 1997; Table 1), which is a common pathological lesion in older mice (Bronson and Lipman 1991). On the basis of this finding it is tempting to suggest that the higher mutant frequencies in liver predict an organ-specific predisposition to cancer.

In the corresponding human syndrome, xeroderma pigmentosum, internal tumor development is rare, but neurological abnormalities are frequently observed (Cleaver and Kraemer 1995). In *XPA*-deficient mice such abnormalities have thus far not been found. Phenotypical differences between *XPA*-deficient mice and human *XPA* might be due to tissue-dependent variation in the levels of endogenous DNA damaging species. This may lead in humans to deleterious effects in brain, whereas in mice it may cause a higher frequency of spontaneous liver tumors. Humans with XP rarely survive beyond the third decade of life, a consequence of the dramatic increase in sunlight-induced skin cancer (Cleaver and Kraemer 1995). Skin cancer does not occur in rodents, which have a fur that cannot be penetrated by UV and are also kept under conditions not permitting exposure to sunlight. This explains the lack of such a phenotype of the *XPA* mutation at early ages. In the young animals NER could be essentially redundant, but become increasingly important at later ages. Indeed, it has been repeatedly argued that loss of re-

Table 1. Spontaneous tumors arising in aging $XPA^{-/-}$ and control mice (≥ 1 year)[a]. (de Vries et al. 1997)

Age range (months)	Genotype	No. of tumors/ no. of animals analyzed	Tumor type
12–14	Control	0/3	n.a.
	$XPA^{-/-}$	0/5	n.a
15–16	Control	n.d.	n.a.
	$XPA^{-/-}$	1/8	Hepatocellular adenoma
17–18	Control	0/19	n.a.
	$XPA^{-/-}$	1/5	Lung adenoma
19–20	Control	n.d.	n.a.
	$XPA^{-/-}$	3/5	Hepatocellular adenoma (3×)

[a] The presence of tumors in aging mice of all three genotypes was detected by gross necroscopy followed by histopathological examination. The overall incidence of tumors in $XPA^{-/-}$ mice (≥ 1 year of age) was 21% (5 of 24). No tumors were found in 22 $XPA^{+/-}$ and $XPA^{+/+}$ mice, indicated as control mice. n.a. not applicable; n.d. not determined.

dundancy in, e.g., cell number or gene copy number, could be responsible for the gradual loss of individual stability and increased incidence of disease associated with aging (Strehler and Freeman 1980).

Although a relationship between defects in nucleotide excision repair and accelerated aging has thus far not been established, there is clearly a relationship with cancer. This is suggested not only by the corresponding increases in spontaneous age-related mutation accumulation and tumor formation in the XPA-deficient mouse model, but also by the results from experiments in which mutations were induced. Mutant frequencies and tumor formation were studied in several organs after treatment with benzo[a]pyrene (B[a]P). B[a]P induced so-called bulky adducts that require NER for their removal. As shown in Fig. 6, B[a]P induces mutations in both the XPA-deficient and the control lacZ animals. The highest mutant frequencies were found in the spleen, which also appeared to be the organ in which most of the tumors were found (de Vries et al. 1997). Higher mutant frequencies as well as a higher tumor frequency, as compared to the control animals, were found in the XPA-deficient animals (Fig. 7). In this respect, it can be concluded that XPA deficiency is associated with accelerated B[a]P-induced mutagenesis and tumor formation in the same organ. Interestingly, when in addition to the XPA null genotype one defective TP53 allele was introduced, tumor formation increased even further (Fig. 7). The mutant frequencies, however, were not affected (Fig. 6). These results are in keeping with those presented above, indicating that p53 does not affect mutation induction by genotoxic agents.

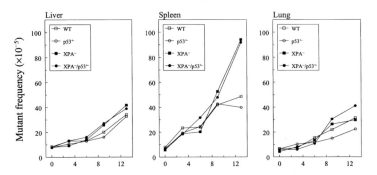

Fig. 6. Average mutant frequencies in liver, spleen and lung of benzo[a]pyrene-treated p53 heterozygote and XPA-deficient lacZ transgenic mice (n = 4). The mice received 13 mg benzo[a]pyrene orally per kg bodyweight, three times per week during 3, 6, 9, or 13 weeks. (After: Effect of heterozygous loss of p53 on benzo[a]pyrene-induced mutations and tumors in DNA repair-deficient XPA transgenic mice, by van Oostrom et al., 1999)

Fig. 7. Percentage of tumor-bearing mice among intercurrent deaths of benzy[a]pyrene-treated mice. The mice received 13 mg benzo[a]pyrene orally per kg bodyweight, three times per week during 13 weeks; data obtained from mice that died intercurrently 52 weeks after start of the treatment. (After: Effect of heterozygous loss of p53 on benzo[a]pyrene-induced mutations and tumors in DNA repair-deficient *XPA* transgenic mice, by van Oostrom et al., 1999)

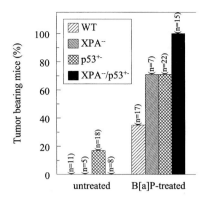

5
Summary and General Discussion

Mutation accumulation in the somatic genome has been implicated in the exponential increase in cancer incidence with age. Using a transgenic mouse model harboring chromosomally integrated *lacZ* reporter genes that can be recovered into *E. coli* and inspected for a wide range of different mutations, we have now demonstrated that mutations accumulate with age at an increased rate in mouse mutants deficient for p53 or *XPA* nucleotide excision repair. The increased rate of mutation accumulation first and foremost points to a relationship with cancer. Both spontaneous and induced mutation accumulation appeared to coincide with tumor formation. Somewhat surprisingly, p53-deficiency had no effect on the mutation frequency at early age or after treatment with genotoxic agents. It does, however, increases the rate of mutation accumulation in vivo at later ages. The *XPA* defect in nucleotide excision repair also appeared to influence the in vivo rate of mutation accumulation during aging and also influenced mutation induction by genotoxic agents. The increased spontaneous mutation accumulation coincided with the reported early appearance of liver tumors in the *XPA*-deficient mouse.

How could somatic mutation loads of the magnitude given above be responsible for the adverse effects associated with the aging process in mammals? First, they could explain the well-known age-related increase in the incidence of cancer. The increased rate of mutation accumulation observed in the $XPA^{-/-}$ animals, which coincided with an increased frequency of tumors in the same organ (liver) underscores this possibility. Second, somatic mutations could be responsible for the loss of cells in various organs, such as the brain. In this respect, empirically determined in vivo mutation frequencies are always underestimates due to the elimination of cells with a high mutation load. This could explain the lack of mutation accumulation in some organs, such as the brain. Finally, random somatic mutations, especially large genome rearrangements, could lead to impaired cell functioning rather than cell death or cell transformation. In this view, the accumulation of random mutations would result in a mosaic of cells at various levels of deficiency. It is conceivable that especially large structural alterations at mutational hotspots could gradually impair genome functioning. Indeed, rather than a catalogue of useful genes interspersed with functionless DNA, each chromosome is now viewed as a complex information organelle with sophisticated maintenance and control systems. Destabilization of these structures by DNA mutations may lead to changes in gene expression, for example, by influencing patterns of DNA methylation and conformation.

Finally, at the current state of technology, a proposed role of somatic DNA alterations in aging is a testable hypothesis. Assuming that the rate of mutation accumulation and the type of mutations that accumulate are determined by the levels of proficiency of the various genome stability systems, identification of the genes involved in these processes should permit the identification of longevity-associated genome stability genotypes. Transgenic mouse modeling of these genotypes and monitoring of genome stability using chromosomally integrated reporter genes, as described in this chapter, will then allow for the direct testing of these potential molecular determinants of aging for their phenotypic consequences, in terms of markers for cellular senescence, pathophysiological variables and life span. This concept is not new (Martin 1977), but its practical implementation has now become a concrete rather than an abstract possibility.

References

Albertini RJ, Hayes RB (1997) Somatic cell mutations in cancer epidemiology. IARC Sci Publ 142:159–184

Bodnar AG, Ouellette M, Frolkis M, Holt SE, Chiu CP, Morin GB, Harley CB, Shay JW, Lichtsteiner S, Wright WE (1998) Extension of life span by introduction of telomerase into normal human cells. Science 279:349–352

Boerrigter METI, Dollé MET, Martus H-J, Gossen JA, Vijg J (1995) Plasmid-based transgenic mouse model for studying in vivo mutations. Nature 377:657–659

Bootsma D, Kraemer KH, Cleaver J, Hoeijmakers JHJ (1998) Nucleotide excision repair syndromes: xeroderma pigmentosum, Cockayne syndrome and trichothiodystrophy. In: Vogelstein B, Kinzler KW (eds) Genetic basis of human cancer, Chapt 13. McGraw-Hill, New York, pp 245–274

Bronson RT, Lipman RD (1991) Reduction in rate of occurrence of age-related lesions in dietary restricted laboratory mice. Growth Dev Aging 55:169–184

Burnet FM (1974) Intrinsic mutagenesis: a genetic approach to aging. Wiley, New York

Campisi J (1996) Replicative senescence: an old lives' tale? Cell 84:497–500

Cleaver JE, Kraemer KH (1995) In: Scriver CR, Beaudet AL, Sly WS, Valle D (eds) The metabolic and molecular bases of inherited disease. 7th edn., vol III. McGraw-Hill New York, pp 4393–4419

Curtis H (1963) Biological mechanisms underlying the aging process. Science 141:686–694

De Vries A, Van Oostrom CThM, Hofhuis FMA, Dortant PM, Berg RJW, de Gruijl FR, Wester PW, Van Kreijl CF, Capel PJA, Van Steeg H, Verbeek SJ (1995) Increased susceptibility to ultraviolet-B and carcinogens of mice lacking the DNA excision repair gene XPA. Nature 377:169–173

De Vries A, Van Oostrom CThM, Dortant PM, Beems RB, Van Kreijl CF, Capel PJA, Van Steeg H (1997) Spontaneous liver tumours and benzo[a]pyrene-induced lymphomas in XPA-deficient mice. Mol Carcinogen 19:46–53

Dollé MET, Giese H, Hopkins CL, Martus H-J, Hausdorff JM, Vijg J (1997) Rapid accumulation of genome rearrangements in liver but not in brain of old mice. Nature Genet 17:431–434

Failla G (1958) The aging process and carcinogenesis. Ann NY Acad Sci 71:1124–1135

Fearon ER, Vogelstein B (1990) A genetic model for colorectal tumorigenesis. Cell 61: 759–767

Friedberg EC, Walker GC, Siede W (1995) DNA repair and mutagenesis. ASM Press, Washington DC

Fukuchi K, Martin GM, Monnat RJ Jr (1989) Mutator phenotype of Werner syndrome is characterized by extensive deletions. Proc Natl Acad Sci USA 86:5893–5897

Giese H, Dollé MET, Hezel A, van Steeg H, Vijg J (1999) Accelerated accumulation of somatic mutations in mice deficient in the nucleotide excision repair gene XPA. Oncogene 18:1257–1260

Gossen JA, Vijg J (1993a) A selective system for LacZ⁻ phage using a galactose-sensitive E. coli host. Biotechniques 14:326–330

Gossen JA, Vijg J (1993b) Transgenic mice as model systems for studying gene mutations in vivo. Trends Genet 9:27–31

Gossen JA, de Leeuw WJF, Tan CHT, Lohman PHM, Berends F, Knook DL, Zwarthoff EC, Vijg J (1989) Efficient rescue of integrated shuttle vectors from transgenic mice: a model for studying gene mutations in vivo. Proc Natl Acad Sci USA 86:7971–7975

Gossen JA, de Leeuw WJF, Molijn AC, Vijg J (1993) Plasmid rescue from transgenic mouse DNA using LacI repressor protein conjugated to magnetic beads. Biotechniques 14:624–629

Gossen JA, Martus HJ, Wei JY, Vijg J (1995) Spontaneous and X-ray-induced deletion mutations in a lacZ plasmid-based transgenic mouse model. Mutat Res 331:89–97

Greenblatt MS, Bennett WP, Hollstein M, Harris CC (1994) Mutations in the p53 tumor suppressor gene: clues to cancer etiology and molecular pathogenesis. Cancer Res 54:4855–4878

Gu H, Marth JD, Orban PC, Mossmann H, Rajewsky K (1994) Deletion of a DNA polymerase beta gene segment in T cells using cell type-specific gene targeting. Science 265:103–106

Hart RW, D'Ambrosio SM, Ng KJ, Modak SP (1979) Longevity, stability and DNA repair. Mech Aging Dev 9:203–223

Hartwell LH, Kastan MB (1994) Cell cycle control and cancer. Science 266:1821–1827

Hoeijmakers JHJ (1994) Human nucleotide excision repair syndromes: molecular clues to unexpected intricacies. Eur J Cancer 30A:1912–1921

Jacks T, Remington L, Williams BO, Schmitt EM, Halachmi S, Bronson RT, Weinberg RA (1994) Tumor spectrum analysis in p53-mutant mice. Curr Biol 4:1–7

Jackson AL, Chen R, Loeb LA (1998) Induction of microsatellite instability by oxidative DNA damage. Proc Natl Acad Sci USA 95:12468–12473

Kohler SW, Provost GS, Fieck A, Kretz PL, Bullock WO, Sorge JA, Putman DL, Short JM (1991) Spectra of spontaneous and mutagen-induced mutations in the LacI gene in transgenic mice. Proc Natl Acad Sci USA 88:7958–7962

Lengauer C, Kinzler KW, Vogelstein B (1998) Genetic instabilities in human cancers. Nature 396:643–649

Levine AJ (1997) p53, the cellular gatekeeper for growth and division. Cell 88:323–331

Loeb LA (1991) Mutator phenotype may be required for multistage carcinogenesis. Cancer Res 51:3075–3079

Lowe SW, Schmitt SW, Smith SW, Osborne BA, Jacks T (1993) p53 is required for radiation-induced apoptosis in mouse thymocytes. Nature 362:847–849

Martin GM (1977) Cellular aging – postreplicative cells. A review (part II) Am J Pathol 89: 513–530

Martin GM (1991) Genetic and environmental modulations of chromosomal stability: their roles in aging and oncogenesis. Ann NY Acad Sci 621:401–417

Murakami Y, Sekiya T (1998) Accumulation of genetic alterations and their significance in each primary human cancer and cell line. Mutat Res 400:421–437

Peto R, Roe FJ, Lee PN, Levy L, Clack J (1975) Cancer and aging in mice and men. Br J Cancer 32:411–426

Ramsey MJ, Moore II DH, Briner JF, Lee DA, Olsen LA, Senft JR, Tucker JD (1995) The effects of age and life style factors on the accumulation of cytogenetic damage as measured by chromosome painting. Mutat Res 338:95–106

Rudolph KL, Chang S, Lee HW, Blasco M, Gottlieb GJ, Greider C, DePinho RA (1999) Longevity, stress response, and cancer in aging telomerase-deficient mice. Cell 96:701–712

Shiloh Y, Rotman G (1996) Ataxia-telangiectasia and the ATM gene: linking neurodegeneration, immunodeficiency, and cancer to cell cycle checkpoints. J Clin Immunol 16:254–260

Skopek TR, Kort KL, Marino DR (1995) Relative sensitivity of the endogenous hprt gene and lacI transgene in ENU-treated Big Blue B6C3F1 mice. Environ Mol Mutagen 26:9–15

Smith ML, Fornace AJ Jr (1995) Genomic instability and the role of p53 mutations in cancer cells. Curr Opin Oncol 7:69–75

Strehler BL, Freeman MR (1980) Randomness, redundancy and repair: roles and relevance to biological aging. Mech Aging Dev 14:15–38

Sugasawa K, Ng JM, Masutani C, Iwai S, van der Spek PJ, Eker AP, Hanaoka F, Bootsma D, Hoeijmakers JH (1998) Xeroderma pigmentosum group C protein complex is the initiator of global genome nucleotide excision repair. Mol Cell 2:223–232

Szilard L (1959) On the nature of the aging process. Proc Natl Acad Sci USA 45:35–45

Tao KS, Urlando C, Heddle JA (1993) Comparison of somatic mutation in a transgenic versus host locus. Proc Natl Acad Sci USA 90:10681–10685

Van Gool AJ, Van der Horst GTJ, Citterio E, Hoeijmakers JHJ (1997) Cockayne syndrome: defective repair of transcription? EMBO J 16:4155–4162

Van Oostrom CThM, Boeve M, van den Berg J, de Vries A, Dollé MET, Beems RB, van Kreijl CF, Vijg J, van Steeg H (1999) Effect of heterozygous loss of p53 on benzo[a]pyrene-induced mutations and tumors in DNA repair-deficient XPA mice. Environ Mol Mutagen 34:124–130

Vijg J, van Steeg H (1998) Transgenic assays for mutations and cancer: current status and future perspectives. Mutat Res 400:337–354

Yu C-E, Oshima J, Fu Y-H, Wijsman EM, Hisama F, Alisch R, et al. (1996) Positional cloning of the Werner's syndrome gene. Science 272:258–262

Delayed Aging in Ames Dwarf Mice.
Relationships to Endocrine Function and Body Size

Andrzej Bartke

1
Introduction

There is considerable evidence for genetic control of aging and longevity. For example, there are major differences in life span between different breeds of domestic dogs, between different inbred lines of laboratory mice, and between lines of mice developed by selection for various phenotypic traits. Studies in yeast, in a round worm, *Caenorhabditis elegans*, and in the fruit fly, *Drosophila melanogaster*, provided a wealth of information on the genetic control of aging, including evidence that mutations at specific loci can produce major alterations in the life span. A new term, longevity assurance genes (LAGs), was coined to describe genes involved in the control of the life span. Evidence for the existence of LAGs in yeast, *C. elegans*, and *Drosophila* and the effects of mutations at these loci are described in other chapters in this Volume.

Although existence of LAGs in higher organisms can be suspected from these observ., there are surprisingly few examples of mutations at specific loci affecting aging in mammals or in the human. In the laboratory mouse, mutations at many loci and targeted disruption (knockout) of numerous genes are associated with significant reduction of life expectancy. However, premature death of the affected animals appears to be due to developmental abnormalities or identifiable functional deficits rather than to accelerated aging. There is recent evidence for genetic basis of at least one of several progeria (premature aging) syndromes in the human (Ye et al. 1998) and a mutation causing a syndrome resembling aging was recently described in the mouse (Yamashita et al. 1998). However, it remains to be determined whether these syndromes can be viewed as premature onset and rapid progression of otherwise normal aging or should be regarded as degenerative diseases with symptoms overlapping those that accompany aging. It is against this background that the question of existence of genes that can extend the life span of

Department of Physiology, Southern Illinois University School of Medicine, Carbondale, IL 62901-6512, USA

Results and Problems in Cell Differentiation, Vol. 29
Hekimi (Ed.): The Molecular Genetics of Aging
© Springer-Verlag Berlin Heidelberg 2000

a mammal assumes special significance. One example of delayed aging in mammals caused by mutation at a single locus was provided by our report of markedly increased life span in Ames dwarf mice (Brown-Borg et al. 1996). Comparable extension of life span was demonstrated also in Snell dwarf mice, which are homozygous for a different mutation but phenotypically are very similar, if not identical, to the Ames dwarfs (K. Flurkey and D. Harrison, per. comm., R. Miller, per. comm.). Thus, it would appear that two loci, Prop-1 for the Ames dwarfs and Pit-1 for the Snell dwarfs, may be regarded as the first identified LAGs in a mammal. In this chapter, we will describe the phenotypic characteristics of Ames and Snell dwarf mice, review information on the life span and age-associated changes in these animals, and propose mechanisms that may be responsible for their prolonged survival.

2
Ames Dwarf Mice

Ames dwarf mice are homozygous for a recessive mutation which was discovered in the breeding colony of laboratory mice at Iowa State University in Ames, Iowa. This mutation was originally described by Schaible and Gowen (1961) and named Ames dwarf (genetic symbol, df). The DF locus was mapped to chromosome 7 (currently designated chromosome 11) by analyzing recombination frequency with Rex and waved-2, which were known to be located on this chromosome (Bartke 1965a). More recent studies based on positional cloning resulted in precise mapping of this gene on chromosome 11 (Buckwalter et al. 1991; Sornson et al. 1996). Mice homozygous for Ames dwarfism (df/df) are phenotypically normal at birth, but their growth and development are retarded. Starting at the age of approximately 14 days, Ames dwarf mice are significantly smaller than their normal siblings (df/+ and +/+, which apparently do not differ from each other). Their postweaning growth rate is markedly reduced and adult Ames dwarf mice are diminutive, weighing approximately 1/3 of the weight of their normal siblings (Fig. 1). In addition to reduced body size, Ames dwarfs are distinguished from normal mice by somewhat infantile body proportions, with a short snout being perhaps the most striking feature. Although less active than normal mice, the Ames dwarfs appear healthy and, under conditions of our animal colony, outlive their normal siblings by an average of at least 1 year.

The mechanisms responsible for retarded development, diminutive body size, and infertility of Ames dwarf mice are well understood. In most elegant studies, Sornson and his colleagues (Sornson et al. 1996) demonstrated that this mutation prevents differentiation of specific cell lineages in the anterior pituitary on embryonic day 12.5–13 (Fig. 2). Since the affected cell types would normally express the homeotic factor Pit-1, these investigators named

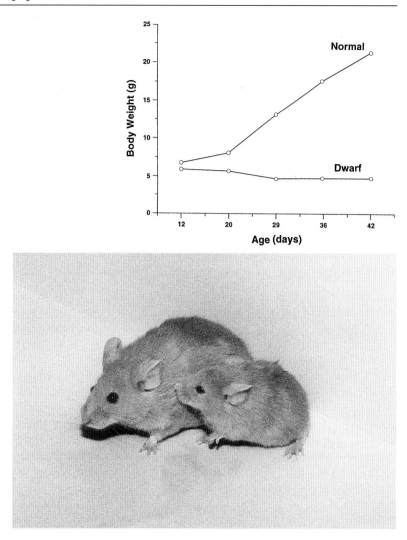

Fig. 1. *Top.* Body weight of Ames dwarf (df/df) and normal (+/?) mice. Each *point* represents the mean value of five to seven animals. (After Romero and Phelps 1993). *Bottom.* Adult Ames dwarf mouse (*right*) and its normal sibling (*left*). (Photo courtesy Dr. W. Hunter)

the Ames dwarf locus Prophet of Pit-1, abbreviated Prop-1. As the result of developmental arrest in the pituitary of the Ames dwarf mouse, the following three cell types fail to differentiate: (1) the somatotrophs which normally produce growth hormone (GH), (2) the lactotrophs, which normally produce prolactin (PRL), and (3) the thyrotrophs which normally produce thyroid-stimulating hormone (TSH). Consequently, Ames dwarf mice are GH-, PRL-, and TSH-deficient. This has been amply documented by histological and

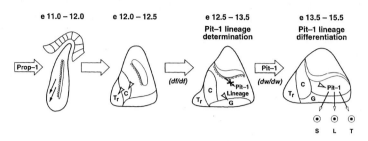

Fig. 2. Model of determination and differentiation events in pituitary ontogeny. The determination of corticotrope (*C*), rostral tip thyrotrope (*Tr*), and gonadotrope (*G*) cell phenotype between embryonic days (*e*) 8.5 and 11.5, and their subsequent differentiation occur before determination of the Pit-1 lineages on ~e12.5–13.0, with Pit-1 initially detected on e13.5. Determination/migration is denoted by *small arrows*, differentiation by *open arrowheads*. The Prop-1 mutation in df/df results in a failure of determination of the Pit-1 lineages (e12.5-e13.0). *Cross* determination block in df/df mouse; *Pit-1* gene activation and initial differentiation is followed by the appearance of three distinct cell types: the somatotrope (*S*), lactotrope (*L*), and thyrotrope (*T*) cell types. (After Sornson et al.)

immunocytochemical studies of the pituitary (Bartke 1964; Andersen et al. 1995), radioimmunoassay measurements of hormone levels in peripheral circulation (Barkley et al. 1982), and results of hormone replacement studies (Bartke 1965b; Doherty et al. 1980; Soares et al. 1984; Chandrashekar and Bartke 1993; Andersen et al. 1995). Most of the phenotypic characteristics of Ames dwarf mice represent the expected consequences of complete, life-long deficiency of GH, PRL, and TSH. Thus, GH deficiency leads to dramatically reduced levels of insulin-like growth factor I (IGF-I, the key mediator of GH action) in peripheral plasma (Chandrashekar and Bartke 1993), reduced growth, and diminutive body size. Deficiency of PRL accounts for luteal failure and consequent sterility of Ames dwarf females (Bartke 1965c, 1979a). In the mouse, PRL is an essential component of the hormonal complex controlling production of progesterone by the corpora lutea of the ovary and maintenance of early pregnancy (Bartke 1965c, 1973). Deficiency of TSH leads to hypothyroidism (Bartke 1964; K. E. Borg and A. Bartke, unpubl. observ.) and persistence of infantile body proportions in the adult. Ames dwarf mice have been used extensively as an animal model for defining the physiological role of adenohypophyseal hormones. Results of these studies included the demonstration that development and functional activity of tuberoinfundibular dopaminergic neurons in the hypothalamus are PRL-dependent (Romero and Phelps 1993), and that both PRL and GH can stimulate testicular function (Bartke 1966a, 1979a; Chandrashekar and Bartke 1993).

Phenotypic characteristics of Ames dwarf mice which are of particular interest in the context of their prolonged longevity are those that resemble the effects of caloric restriction. Caloric restriction can delay aging and prolong life span of mice (Duffy et al. 1990; Masoro 1995; Weindruch and Sohal 1997). These characteristics include reduced body size and delayed

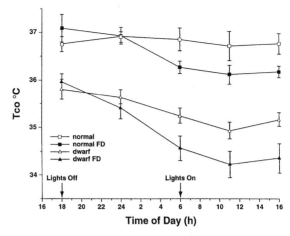

Fig. 3. Body core temperature (*Tco*) of Ames dwarf mice and their normal siblings during two consecutive 24-h periods. Each *point* represents mean data from six mice averaged over the preceding time periods (2, 6, or 5 h) ± SEM. *Open squares* (normals) and *open triangles* (dwarfs) indicate data collected during 24-h baseline period and *solid squares* (normals) and *solid triangles* (dwarfs) represent data collected during the succeeding 24-h food deprivation period. Please note similar Tco responses to food deprivation in dwarf and normal mice. (After Hunter et al. 1999)

puberty, reduction in core body temperature (Bartke et al. 1998; Hunter et al. 1999; Fig. 3), reduction in plasma IGF-I and glucose, reduction in insulin levels in females, and increase in corticosterone levels in males (Borg et al. 1995). However, under standard housing conditions, i.e., constant access to food, dwarf mice consume more food per gram body weight than their normal siblings (Mattison and Bartke, unpubl.), and frequently become obese. Thus, longevity of dwarf mice does not appear to be due to "voluntary" caloric restriction. These findings will be discussed in some detail later in the chapter. Counter-intuitive in the context of delayed aging are several indices of reduced immune function in Ames dwarf mice (Duquesnoy 1972; Esquifino et al. 1991).

3
Snell Dwarf Mice

In 1929, Snell (1929) reported the existence of a recessive mutation which caused dwarfism in mice, and named it dwarf (genetic symbol dw). Deficiency of pituitary function in dw/dw dwarf mice was demonstrated in a series of developmental and hormone replacement studies (review in Grüneberg 1952). Demonstration that this mutation causes primary defect in adenohypophyseal rather than hypothalamic function was obtained in studies involving reciprocal pituitary transplantation between dwarf and normal mice (Carsner and Rennels 1960). More recent studies demonstrated various secondary alterations in the hypothalamus, some of which could be corrected by appropriate replacement with anterior pituitary hormones (Morgan et al. 1981; Webb et al. 1985; Phelps 1994). The DW locus is located on the chromosome 16 and thus is unrelated to DF. In 1990 (Li et al. 1990), it was shown that dwarf mice, which in the meantime were renamed Snell dwarf mice, are homozygous for a mutation at the Pit-1 locus. As was mentioned earlier in this chapter, Pit-1 is involved in the differentiation of somatotrophs, lactotrophs, and thyrotrophs on embryonic days 13.5–15.5 (Sornson et al. 1996). It binds to the promoters of the GH, PRL, and TSHβ genes and stimulates their expression (Schaufele 1994; Holloway et al. 1995). Thus, Pit-1 mutants, similarly to Prop-1 mutants, lack somatotrophs, lactotrophs, and thyrotrophs and are GH-, PRL-, and TSH-deficient. In other words, phenotypes of Snell dwarf mice are very similar, if not identical, to the phenotypes of Ames dwarfs even though the corresponding mutations are genetically unrelated and located on different chromosomes. Quantitative differences in various traits of Snell and Ames dwarf mice as well as quantitative differences between findings of different investigators studying the same mutant are almost certainly due to differences in genetic background. For example, gonadal development of Snell dwarf mice is more advanced in animals on a heterogeneous genetic background than in those derived from an inbred strain (Bartke and Lloyd 1970). Although both Snell and Ames dwarf mice reach approximately 30–35% of the body weight of their normal siblings, the absolute body weight of dwarf animals can differ substantially, in proportion to differences in body weight between normal animals from the corresponding lines (Bartke and Lloyd 1970; A. Bartke, unpubl. observ.).

Snell dwarf mice were used extensively as a model of congenital GH deficiency and hypothyroidism (Grüneberg 1952). Evidence that they are also PRL-deficient became available only 35 years after their initial description (Bartke 1965b; Bartke 1966a,b). They continue to be a popular animal model for the study of the GH-IGF-I axis, immune function, etc.; their phenotype has been characterized in considerable detail. The characteristics of Snell dwarf mice which could have some relevance to their delayed aging include small body size, low metabolic rate and body temperature (Boettiger et al. 1940), reduced number of mitotic divisions during development leading to reduced cell number in the various organs (Viola-Magni 1965; Winick and Grant 1968), hypogonadism (Grüneberg 1952; Bartke and Lloyd 1970; Bartke 1979b), hypothyroidism (Grüneberg 1952; Lewinski et al. 1984), and extremely low levels of IGF-I (Holder et al. 1980; van Buul Offers et al. 1986). Similarly to Ames dwarf mice, immune function in the Snell dwarfs appears to be compromised (Duquesnoy et al. 1970; Fabris et al. 1972; Pelletier et al. 1976).

4
Development and Longevity of Dwarf Mice

As was mentioned earlier in this chapter, the Ames dwarf mice, in comparison to their normal siblings, are much smaller, have infantile body proportions, are less active, and retarded in their sexual development. In our colony with heterogeneous genetic background derived from crosses of several lines, male dwarf mice undergo pubertal development with obvious testicular growth and scrotal descent and, similarly to normal males, exhibit aggressive behavior toward nonfamiliar male dwarfs. They have complete spermatogenesis and attain testicular size which is smaller than normal in terms of absolute weight but larger than normal when expressed per gram body weight (Bartke 1979a; Chandrashekar and Bartke 1993). However, most males are infertile and only an occasional animal will sire a few litters if housed with very young (i.e., small) normal females. Sexual maturation of females is retarded, as judged by the age of establishment of vaginal opening and many females never reach puberty. Sexual maturation can be induced by treatment with thyroxine (T_4), GH, or a combination of T_4 and GH (Bartke 1979b; Soares et al. 1984). Transplantation of pituitary glands from normal females under the renal capsule of the dwarfs can also induce sexual maturation (Bartke 1979b). Most of the Ames dwarf females that achieved sexual maturation spontaneously or as a result of hormonal treatment will mate but never become pregnant. Typically, the matings will occur at regular 4–6-day intervals corresponding to the length of the estrous cycle in normal female mice. This indicates that the animals have ovulatory cycles of approximately normal length, but they fail to respond to stimuli associated with mating by activation of the corpora lutea. In normal female mice, mating induces twice daily surges of PRL (Barkley et al. 1978) which lead to persistence of corpora lutea and maintain pregnancy (Bartke 1973). Sterile matings in normal animals are similarly followed by induction of PRL surges and thus lead to luteal activation, the so-called pseudopregnancy. Pseudopregnancy delays the next ovulation and mating for 10–11 days (Bartke et al. 1972). In Ames dwarf females, pregnancy can be maintained by treatment with injections of PRL or with ectopic transplants of normal pituitaries which are known to secrete PRL (Bartke 1979a). Dwarf females with a normal pituitary transplanted under the renal capsule can produce a series of litters and raise their pups to weaning (Bartke 1965c, 1979a). Collectively, these data indicate that luteal failure and infertility of Ames dwarf females are due to PRL deficiency.

Our study of longevity of Ames dwarf mice was prompted by chance observ. that these animals appear to remain in excellent general condition longer than their normal siblings. They tend to become obese (which is also common in the normal animals from this line) and some experience hair loss

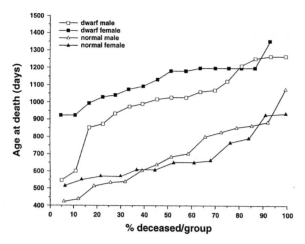

Fig. 4. Longevity in male and female normal and Ames dwarf mice. Each *point* on the graph represents an individual animal surviving to the specific age indicated versus the percentage of animals deceased per group. *Open symbols* males; *closed symbols* females; *triangles* normals; *squares* dwarfs. Dwarfs live longer than normal mice regardless of gender ($P < 0.0001$). (Brown-Borg et al. 1996)

but otherwise appear very healthy. When a group of 34 dwarf and 28 normal mice were allowed to live their natural life span, the dwarfs outlived the normal mice by more than a year (Fig. 4). The difference in the average life span was 353 days for males and 488 days for females (Brown-Borg et al. 1996). This corresponds to extension of the life span by 49% in males and by the remarkable 64% in females. Two of the female dwarfs lived 4 years and 2 months. Most of the dwarfs remained in very good body condition well past the average life span of their normal siblings and exhibited age-related decline (weight loss, sluggishness, occasional cataracts) only toward the end of their life. A study of locomotor activity revealed that 3-year-old Ames dwarf mice did not exhibit any evidence of age-related reduction and, in fact, were more active than young dwarfs in terms of both ambulatory and stereotyped behaviors (Meliska and Bartke, unpubl. observ.). Normal animals examined in the same study were more active than dwarfs. However, at the age of approximately 20 months, their ambulatory and stereotyped movements tended to decline rather than increase, in comparison to the values measured in young normal mice (Meliska and Bartke, unpubl. observ.).

We are currently studying histopathological changes in aging Ames dwarf and control (normal) mice. Results available to date indicate that incidence of lung tumors and liver tumors as well as incidence of all types of tumors in Ames dwarf mice dying of natural causes are comparable to the incidence of these lesions in normal animals dying from natural causes. Since dwarf mice live longer, there is reason to suspect that they develop tumors at a more advanced chronological age, and additional studies to test the validity of this interpretation are indicated. Data on tumor incidence in aging Ames dwarfs imply that age-related pathological changes in these animals are not prevented or altered although they are likely to be delayed. This suggests yet another similarity between the effects of hereditary dwarfism and effects of caloric restriction and argues in favor of considering the Ames dwarf mice as a novel model system to study delayed but otherwise normal aging.

The observed incidence of tumors in aging Ames dwarf mice was not anticipated from the reports of "tumor resistance" in Snell dwarf mice. Snell dwarf mice were reported to develop fewer spontaneous or chemically induced tumors than the normal animals (Bielschowsky and Bielschowsky 1961; Chen et al. 1972) and to exhibit reduced growth of transplanted tumors (Rennels et al. 1965). The natural history of tumors in dwarf mice is of particular interest in the context of the recent evidence that IGF-I is an important cancer risk factor and that it can promote growth of different types of tumors (Werner and LeRoith 1996).

5
Longevity of Snell Dwarf Mice and the Issues of Husbandry

An early report (Fabris et al. 1972) claimed that Snell dwarf mice are extremely short-living with an average life span of 4–5 months. This was ascribed to reduced immune competence of these animals and the consequent susceptibility to infectious disease (Fabris et al. 1972). These findings must have been unique to the particular population of mice used for these studies or, more likely, to the husbandry conditions, because soon after their publication, three different laboratories responded by reporting that Snell dwarf mice in their colonies live at least as long as normal animals (Shire 1973; Schneider 1976; Bartke 1977). However, these investigators did not report how long their animals live, and this problem was apparently not systematically investigated until the mid 1990s. Since that time, it was determined in two laboratories that Snell dwarf mice live significantly longer than normal animals from the same strain (R. Flurkey, D. Harrison, and R. Miller, pers. comm.). We believe that the discrepancies between data on longevity of Snell dwarf mice in different laboratories are almost certainly due to differences in husbandry and to the general improvement in hygiene and animal health standards in the animal colonies during the past 50 years. Although Snell dwarf mice are immunodeficient (Duquesnoy et al. 1970; Fabris et al. 1972), this may be of little consequence for survival of the animals in modern facilities which are free of most, if not all, of the common murine pathogens.

Since both Snell and Ames dwarf mice have reduced body temperature (Schonholz and Osborn 1949; Hunter et al. 1999) and limited ability for thermoregulation at low ambient temperatures (Grüneberg 1952), their survival can probably be influenced by temperature and housing conditions, e.g., single vs. group housing. Interactions among the group-housed animals can also be important. For example, normal adult males persist in trying to mate with dwarfs of both sexes and apparently can also be aggressive toward these animals. In contrast, housing dwarfs with normal females or with several other dwarfs appears to promote their well being, presumably by helping them to keep warm by huddling with their cage mates. Survival of very young dwarfs presents additional problems since at the usual age of weaning they are tiny and appear to have difficulties in eating food pellets placed in the food hopper or cage lid. Our standard practice is to wean the dwarfs at the age of 6–8 weeks, i.e., 1 month after weaning their normal siblings or at the time of weaning normal animals from the next litter, whichever comes first. After weaning, the dwarfs are housed with other dwarfs or with normal females, in groups of four to five animals in cages equipped with filter (micro-isolator) tops. After weaning, regrouping of the animals is kept to absolute minimum. To avoid fighting, male dwarf mice that were housed since weaning in different cages are never combined. The temperature in the room is maintained at 22 ± 2°C, the light cycle is 12:12, and food and water are available ad libitum.

Although no systematic studies of the effects of housing conditions on the survival of Snell or Ames dwarf mice appeared to have been conducted and there is little published information on their husbandry, it can be assumed that these mutants will not thrive when single-housed or exposed to low ambient temperatures, particularly in wire-bottom cages. Weaning the dwarfs at the age of 3 weeks is likely to interfere with their health and survival unless special arrangements are made to improve food accessibility and palatability and to prevent excessive heat loss. After the age of 1 1/2–2 months, dwarf mice thrive under standard conditions when housed with normal females or with other dwarfs.

6
Possible Mechanisms of Delayed Aging in Dwarf Mice

The remarkable longevity advantage of Ames dwarf mice raises an obvious question, namely: which characteristics of these animals are responsible for prolongation of their life span? Information available to date and comparison with data obtained in other models of delayed aging suggest a number of possibilities which will be briefly discussed below.

6.1
Reduced Blood Glucose and Increased Sensitivity to Insulin

As was mentioned earlier in this chapter, plasma glucose levels in Ames dwarf mice are lower than in the corresponding normal controls, while plasma insulin is normal in female dwarfs and reduced in male dwarfs (Borg et al.

1995). Thus, these animals appear to have enhanced sensitivity to the actions of insulin. Concomitant reduction in glucose and insulin levels is consistently observed in genetically normal animals subjected to caloric restriction (Duffy et al. 1990; Masoro 1995; Weindruch and Sohal 1997). Mice with targeted disruption knockout of the GH receptor gene have extremely low plasma insulin levels and reduced levels of blood glucose (N. Danilovich, J. Mattison, and A. Bartke, unpubl. observ.), and preliminary data suggest that they may live longer than normal animals. Nonenzymatic glycation of proteins and other components of living cells is believed to represent an important mechanism of aging and links elevated blood glucose levels with reduced life expectancy (Reiser 1998). In the human, diabetes and insulin resistance are important risk factors for a number of diseases and for reduced life span. Transgenic mice overexpressing GH are insulin-resistant (Balbis et al. 1996; Dominici et al. 1998), exhibit various symptoms of accelerated aging (Pendergast et al. 1993; Steger et al. 1993; Miller et al. 1995), and have a drastically reduced life span (Wolf et al. 1993; Cecim et al. 1994; Rollo et al. 1996). The possibility of functional and evolutionarily conserved role of energy metabolism and insulin signaling in the control of aging is suggested by the demonstration that a "longevity gene" in a worm, *Caenorhabditis elegans*, is related to the insulin receptor gene in higher organisms (Kimura et al. 1997).

6.2
Hypothyroidism

Both Ames and Snell dwarf mice are severely hypothyroid due to primary TSH deficiency (Grüneberg 1952; Bartke 1964; Li et al. 1990; Sornson et al. 1996). In normal animals, caloric restriction reduces thyroid hormone levels (Meites 1993; Weindruch and Sohal 1997). In rats, experimentally induced hypothyroidism is associated with increased longevity (Denckla 1974; Ooka et al. 1983), while hyperthyroidism is associated with shortened life span (Ooka and Shinkai 1986). Beneficial effects of hypothyroidism on the life span could be related to the role of thyroid hormones in the control of body temperature, metabolic rate, and thus generation of oxygen radicals.

6.3
Reduced Body Temperature and Metabolic Rate

Core body temperature in Ames dwarf mice is reduced by approximately 1.5°C (Hunter et al. 1999). This suggests that metabolic rate is also reduced. Reduced metabolic rate in Snell dwarf mice was described in several studies dating back to the 1930s (Benedict and Lee 1936; review in Grüneberg 1952). In normal animals, caloric restriction reduces body temperature (Lane et al. 1996). Metabolic rate is initially reduced in CR animals but thereafter stabilizes at a level appropriate to the reduction in lean body mass (Duffy et al. 1989, 1990). It is widely accepted that generation of free oxygen radicals (often referred to as reactive oxygen species, ROS) in the course of mito- chondrial energy metabolism represents an important, if not the key mech- anism of aging. Therefore, it can be assumed that reductions in metabolic rate and in body temperature can prolong life by reducing exposure of the cells to ROS.

6.4
Improved Capacity to Remove Reactive Oxygen Species

Oxidative damage, believed to represent a major mechanism of aging, is related not only to the rate of generation of ROS, but also to the rate of their removal (Cutler 1985). Removal of ROS depends on the actions of several enzymes including catalase and glutathione peroxidase, and we have recently shown that the activity of catalase is significantly greater in the liver and the kidney of Ames dwarf mice than in the corresponding organs of normal animals (Bartke et al. 1998). In the liver, the activity of glutathione peroxidase was also greater in dwarf than in normal mice (H. M. Brown-Borg et al. unpubl. data).

These findings suggest that the tissues of dwarf mice may be less vulner- able to oxidative damage by virtue of increased ability to detoxify free oxygen radicals. Consistent with this interpretation, the levels of inorganic peroxides in the liver were found to be lower in dwarf than in normal mice (Bartke et al. 1998). The possible link between the activity of enzymes involved in scav- enging ROS and longevity is supported by evidence for reduced activity of these enzymes and for the increase in oxidative processes in GH transgenic mice which have greatly reduced life span (Rollo et al. 1996; H. M. Brown- Borg, pers. comm.).

6.5
Hypogonadism

In dwarf mice, puberty is delayed and gonadal function is suppressed (re- views in Bartke 1979b, 1982). Production of testosterone by the testes is significantly lower than in normal mice (Bartke et al. 1977; Chandrashekar and Bartke 1993). In Snell dwarf mice from an inbred strain, Dwarf/J, the reproductive system is infantile and spermatogenesis is suppressed (Bartke and Lloyd 1970). Most of the Ames dwarf and Snell dwarf females fail to undergo sexual maturation and thus can be assumed to experience estrogen deficiency throughout their life span. Although effects of gonadal function (more specifically, gonadal steroids) on aging and longevity are not well understood, there is evidence for increased life span of castrated, as com- pared to intact animals (Drori and Folman 1976). Sex steroids have anabolic effects and contribute to pubertal growth and thus could influence longevity by increasing metabolic rate and body size. Relationships of growth and body size to longevity are discussed in another section of this chapter (6.7).

In support of the possible role of hypogonadism in delayed aging of dwarf mice, the longevity advantage (defined as extension of the life span in relation to the normal animals) is greater in females than in male dwarfs (Brown-Borg et al. 1996). Female dwarfs are profoundly hypogonadal while gonadal function in males is much less affected.

6.6
Deficiency of GH and IGF-I

In dwarf mice, GH is not produced and the levels of IGF-I in peripheral circulation are profoundly suppressed and remain near or below the levels of detectability by radio immunoassay (Pell and Bates 1992; Chandrashekar and Bartke 1993). There is considerable evidence for a reciprocal relationship between the function of the GH-IGF-I axis and longevity. In normal rats, caloric restriction leads to significant suppression of IGF-I levels and this effect has been proposed as a likely mechanism for delayed onset of age-related diseases (especially tumors) and for increased longevity (Breese et al. 1991). In these animals, plasma GH levels also decline (Meites 1993), although in older CR rats, GH levels were found to be higher than in ad libitum-fed controls because CR prevents age related decline in GH pulsatility (Sonntag et al. 1995). Mechanisms responsible for suppression of IGF-I levels in older CR animals in which pulsatile GH secretion is preserved remain to be clarified. Preliminary data suggest that GH receptor knockout mice which are GH-resistant and consequently IGF-I-deficient live longer than normal mice (Kopchick and Coschigano, pers. comm.). Studies in different breeds of domestic dogs suggested a reciprocal relationship between plasma IGF-I levels and life span (Patronek et al. 1997). In transgenic mice overexpressing GH genes, plasma GH and IGF-I levels are greatly increased (Naar et al. 1991) and life span is reduced (Wolf et al. 1993; Cecim et al. 1994; Bartke et al. 1998). In the human, the syndrome of acromegaly involves elevation of plasma GH and IGF-I levels due to hypersecretion of GH by anterior pituitary tumors and is associated with reduced life expectancy (Bengtsson et al. 1988; Orme et al. 1998).

The possible mechanisms linking activity of the GH-IGF-I axis and longevity include anabolic and thermogenic actions of GH (Strobl and Thomas 1994) and the dominant role of this axis in determining the rate of growth and the adult body size. There is considerable evidence that, within the species, body size is negatively correlated with longevity (details and references in the next section of this chapter).

Mitogenic actions of IGF-I may be specifically related to the reciprocal relationship between IGF-I levels and longevity. In Snell dwarf mice, mitotic divisions appear to cease early in life and the number of cells in the various organs is reduced (review in Bartke 1979b), implying reduced number of cell divisions. The number of cell divisions in cultured cells determines their remaining "replicative potential" and life span (Hayflick 1998), and cells removed from an older individual's bone reduced potential to undergo divisions in culture (Hayflick 1998). Interestingly, reduced replicative potential was detected in GH transgenic mice in which GH levels were elevated and life expectancy reduced (Pendergast et al. 1993).

However, it may be premature to consider delayed aging of Ames and Snell dwarf mice as the expected outcome of deficiency of GH and IGF-I. Although the evidence for reciprocal relationship between the activity of the GH-IGF-I axis and longevity summarized above appears to be compelling, important issues concerning this relationship remain to be clarified. In the human, both congenital and adult onset GH deficiency are associated with serious functional deficits and increased risk for premature death from cardiovascular disease (Sacca et al. 1994; DeBoer et al. 1995; Bates et al. 1996). Although no data on longevity of humans chronically treated with GH are available, GH-deficient patients clearly benefit from GH replacement therapy (Sacca et al. 1994; Pfeifer et al. 1999). Moreover, numerous symptoms of aging resemble symptoms of GH deficiency, presumably represent consequences of age-related decline in GH levels, and can be corrected by GH replacement therapy (Rudman et al. 1990; Corpas et al. 1993). The possible reasons for discrepancy between these findings and the prolonged survival of GH-deficient animals will be discussed below.

6.7
GH-IGF-I Axis, Body Size, and Aging

Although the present understanding of the mechanisms of normal, delayed, or accelerated aging is limited, causal relationships can be tentatively identified from associations of various phenotypic characteristics with alterations of life span in different models. For example, demonstration of reduced core body temperature, blood glucose, and IGF-I levels in both hereditary dwarf mice and CR animals strengthens the argument that these characteristics may be important in increasing longevity. Analysis of findings in these and other models of delayed aging points to the body size as a key correlate and perhaps a key determinant of longevity. Thus, life span in mice can be extended by hereditary dwarfism (Brown-Borg et al. 1996; Flurkey and Harrison, pers. comm.; Miller, pers. comm.), selection for reduced body size (Roberts 1961), or CR which also reduces growth and body size (Duffy et al. 1990). In rats and mice, CR initiated at the time of weaning has greater impact on body size and causes greater extension of life span than is observed in animals subjected to CR after reaching sexual maturation, or at middle age (Weindruch and Sohal 1997). Remarkably, longevity of rats can be extended by hypophysectomy providing that the animals receive thyroid and glucocorticoid replacement (Everitt et al. 1980). Hypophysectomy produces numerous functional deficits which include arrest of growth, and thus leads to reduced body size in comparison to intact controls. Negative correlation of body size and life span within a species applies not only to mice but also to invertebrates (Austadt 1997), domestic dogs (Patronek et al. 1997), and probably also to humans (Samaras and Storms 1992; Micozzi 1993). At the present time, it is unclear whether small body size per se imparts a longevity advantage or provides a marker for some physiological characteristic or characteristics that affect life span. Small body size could favor longevity by increasing efficiency of the cardiovascular system within the confines of a genetically determined, species-specific body plan (Promislow 1993), or by some other mechanisms specifically related to body dimensions, mass/surface ratio, or weight. Relationship of body size to longevity could also reflect correlation of body size with the amount of metabolic energy processed per unit of time, with the amount of ROS-related damage incurred in the process of metabolism and growth, with regulation of glucose metabolism and thus glycation-related damage of proteins, or with other physiological characteristics related to aging. The impressive longevity advantage of GH-deficient mutants and of CR animals in which the GH-IGF-I axis appears to be suppressed can perhaps be reconciled with the detrimental effects of GH deficiency in humans by assuming that the benefits of reduced body size in genetically dwarf or CR mice outweigh the "risks" of GH deficiency on cardiovascular system or immune function. Naturally, these comparisons are complicated by differences between the species and by differences between the environmental conditions of laboratory rodents and human beings. Limited opportunity or need for physical activity, constant access to high energy food, constant ambient temperature, and effective protection from pathogens are characteristic of "standard" conditions in colonies of laboratory rodents and clearly differ from the conditions and challenges of human existence.

It is also unclear whether evidence that GH can prevent or reverse some of the physiological changes associated with aging in the human (Rudman et al. 1990; Pfeifer et al. 1999) or in experimental animals (Sonntag et al. 1985, 1997) can be taken as indication that GH normally acts to prevent rather than to accelerate aging. There are no data on the effects of GH treatment on longevity in these studies; in fact, the initial report of highly beneficial outcome of GH replacement in the elderly (Rudman et al. 1990) was followed by reports of various detrimental side effects of treatment with GH or IGF-I (Borst et al. 1994; Thompson et al. 1995; Papadakis et al. 1996).

The issue of the dose-response relationship of the actions of GH (and possibly also IGF-I) introduces yet another complication to interpretation of the data concerning GH levels and life span. GH actions can be biphasic with regard to the dose (Ultsch and deVos 1993) and, therefore, it is quite likely that the relative GH deficiency occurring naturally in old age may be detrimental and correctable by treatment with low (Toogood and Shalet 1999) or moderate doses of GH, while GH excess is clearly detrimental to survival (Bengtsson et al. 1988; Orme et al. 1998).

7
General Conclusions and Future Directions

The remarkable extension of life span in dwarf mice clearly establishes these animals as a unique and potentially valuable model for the study of delayed aging in a mammal. Availability of a new model system for this type of investigation is very important because, up to now, CR represented the only well-established experimental model for mechanistic studies of aging in mammals. The effects of CR on the life span and on numerous physiological parameters are well documented and CR animals are generally considered as the system of choice and a "gold standard" in experimental gerontology. However, there are inherent limitations in any attempt to identify causal relationships from the study in a single experimental system, and there are obvious limits to useful generalizations from data generated in any particular model. For example, it can be argued that dramatic extension of life span in CR animals is due in large measure to preventing overeating and obesity, which are characteristic of ad libitum-fed laboratory rodents housed in standard (i.e., relatively small) cages with no access to running wheels, opportunities to exercise, or need to expend energy to find food.

In comparison to the vast literature concerning effects of CR in rats, mice, and, more recently, monkeys (Duffy et al. 1990; Lane et al. 1996; Weindruch and Sohal 1997), the amount of information relevant to prolonged survival of Ames and Snell dwarf mice is limited. However, information available to date and comparisons with findings in CR animals already allow some novel and useful, even if tentative, conclusions. For example, we can assume that while the activity of the GH-IGF-I axis, body size, body temperature, and plasma glucose levels may be critical for prolonged survival, leanness, suppression of insulin levels, and elevation of corticosterone levels are probably much less important or not absolutely required.

The fact that the effects of both Prop-1 Pit-1 genes, as well as primary defects in development and endocrine function of Ames and Snell dwarf mice are well understood, should allow linking specific developmental events and specific aspects of neuroendocrine function to the control of aging. Thus, both of these genes are primary candidates for study as longevity assurance genes (LAGs) in mammals.

The major impact of hereditary dwarfism on aging emphasizes the potential importance of other animal models for this field of investigation. For example, preliminary data suggesting increased longevity of GH receptor knockout mice (Coschigano and Kopchick, pers. comm.) raise a very exciting possibility that a single genetic defect in GH signaling may be sufficient to affect life span. If confirmed, these findings would focus attention on the role of the GH-IGF-I axis and body size in aging. However, caution must be exercised in interpreting these preliminary findings because, as of now, there is no evidence for pro-longed survival of little (lit/lit) mice with isolated GH deficiency or in trans-genic mice with GH resistance due to overexpression of an antagonistic GH analogue and, indeed, there is evidence to the contrary (D. Harrison, pers. comm.; Bartke and Kopchick, unpubl. data). However, there is little doubt that mutants, knockouts, and transgenic animals will soon assume a major role in research directed at identifying factors that control mammalian aging.

The relationship of findings in Ames and Snell dwarf mice and other genetic models to the human is of obvious importance. Direct application of findings in Ames or Snell dwarf mice, and, parenthetically, also in the CR animals, to the human is exceedingly unlikely. Hypopituitarism, hypothy-roidism, stunted growth, and starvation certainly do not represent acceptable clinical interventions or public health measures. Mutations homologous to Prop-1 and Pit-1 have been identified in the human (Tatsumi et al. 1992; Pfäffle et al. 1996; Flück et al. 1998), and the affected children obviously require GH and thyroid hormone replacement. No one would seriously question the benefits of these treatments in these or other hypopituitary patients. However, it is most intriguing that patients with multiple pituitary hormone deficiency due to mutation at the PROP-1 locus were recently re-ported to reach very old ages, exceeding the average life expectancy in the general population (Krzisnik et al. 1999).

Comparative studies in Ames and Snell dwarfs, other mutants, GH-R-KO mice, and CR animals, as well as in various KO and transgenic animals yet to be produced, will undoubtedly lead to identification of the physiological characteristics linked to the rate of aging and the mechanisms involved. Information on the importance of some of these factors, e.g., insulin sensi-tivity, IGF-I levels, or tall stature may help identify individuals at risk for premature aging or, more importantly, for premature onset of age-related diseases and suggest appropriate testing and perhaps also interventions.

References

Andersen B, Pearse RV II, Jenne K, Sornson M, Lin S-C, Bartke A, Rosenfeld MG (1995) The Ames dwarf gene is required for Pit-1 gene activation. Dev Biol 172:495–503

Austad SN (1997) Comparative aging and life histories in mammals. Exp Gerontol 32:23–38

Balbis A, Bartke A, Turyn D (1996) Overexpression of bovine growth hormone in transgenic mice is associated with changes in hepatic insulin receptors and in their kinase activity. Life Sci 59:1363–1371

Barkley MS, Bradford GE, Geschwind II (1978) The pattern of plasma prolactin concentration during the first half of mouse gestation. Biol Reprod 19:291–296

Barkley MS, Bartke A, Gross DS, Sinha YN (1982) Prolactin status of hereditary dwarf mice. Endocrinology 110:2088–2096

Bartke A (1964) Histology of the anterior hypophysis, thyroid and gonads of two types of dwarf mice. Anat Rec 149:225–235

Bartke (1965a) Mouse Newsl 32:52–54

Bartke A (1965b) The response of two types of dwarf mice to growth hormone, thyrotropin, and thyroxine. Gen Comp Endocrinol 5:418–426

Bartke A (1965c) Influence of luteotrophin on fertility of dwarf mice. J Reprod Fertil 10:93–103

Bartke A (1966a) Influence of prolactin on male fertility in dwarf mice. J Endocrinol 35:419–420

Bartke A (1966b) Reproduction of female dwarf mice treated with prolactin. J Reprod Fertil 11:203–206

Bartke A (1973) Differential requirement for prolactin during pregnancy in the mouse. Biol Reprod 9:379–383

Bartke A (1977) Low mortality and long survival of dwarf mice (dw/dw). Mouse Newsl 56:63

Bartke A (1979a) Prolactin-deficient mice. In: Alexander NJ (ed) Animal models for research on contraception and fertility. Harper & Row, Hagerstown, pp 360–365

Bartke A (1979b) Genetic models in the study of anterior pituitary hormones. In: Shire JGM (ed) Genetic variation in hormone systems. CRC Press, Boca Raton, pp 113–126

Bartke A (1982) Influence of prolactin on male gonadal function. In: Clauser H, Gautray J-P (eds) Prolactin neurotransmission et fertilité. Masson, Paris, pp 117–126

Bartke A, Lloyd CW (1970) Influence of prolactin and pituitary isografts on spermatogenesis in dwarf mice and hypophysectomized rats. J Endocrinol 46:321–329

Bartke A, Merrill AP, Baker CF (1972) Effects of prostaglandin F2a on pseudopregnancy and pregnancy in mice. Fertil Steril 23:543–547

Bartke A, Goldman BD, Bex F, Dalterio S (1977) Effects of prolactin (PRL) on pituitary and testicular function in mice with hereditary PRL deficiency. Endocrinology 101:1760–1766

Bartke A, Brown-Borg HM, Bode AM, Carlson J, Hunter WS, Bronson RT (1998) Does growth hormone prevent or accelerate aging? In: Bartke A, Falvo R (eds) Experimental gerontology. Proc 3rd Int Symp on Neurobiol and Neuroendocrinol of Aging, vol 33, pp 675–687

Bates AS, Hoff WV, Jones PJ, Clayton RN (1996) The effect of hypopituitarism on life expectancy. J Clin Endocrinol Metab 81:1169–1172

Benedict FG, Lee RC (1936) La production de chaleur de la souris. Etude de plusieurs races de souris. Ann Physiol Physicochim Biol 12:983–1064

Bengtsson B-Å, Edén S, Ernest I, Odén A, Sjögren B (1988) Epidemiology and long-term survival in acromegaly. Acta Med Scand 223:327–335

Bielschowsky F, Bielschowsky M (1961) Carcinogenesis in the pituitary of dwarf mouse. The response to dimethylbenzanthracene applied to the skin. Br J Cancer 15:257–262

Boettiger EG (1940) The relation of oxygen consumption and environmental temperature to the growth of dwarf mice. Am J Physiol 129:312–313

Borg KE, Brown-Borg HM, Bartke A (1995) Assessment of the primary adrenal cortical and pancreatic hormone basal levels in relation to plasma glucose and age in the unstressed Ames dwarf mouse. Proc Soc Exp Biol Med 210:126–133

Borst SE, Millard WJ, Lowenthal DT (1994) Growth hormone, exercise, and aging: the future of therapy for the frail elderly. J Am Geriatr Soc 42:528–535

Breese CR, Ingram RL, Sonntag WE (1991) Influence of age and long-term dietary restriction on plasma insulin-like growth factor-I (IGF-I), IGF-I gene expression, and IGF-I binding proteins. J Gerontol 46:B180–B187

Brown-Borg HM, Borg KE, Meliska CJ, Bartke A (1996) Dwarf mice and the ageing process. Nature 384:33

Buckwalter MS, Katz RW, Camper SA (1991) Localization of the panhypopituitary dwarf mutation (df) on mouse chromosome 11 in an intersubspecific backcross. Genomics 10: 515–526

Carsner RL, Rennels EG (1960) Primary site of gene action in anterior pituitary dwarf mice. Science 131:829

Cecim M, Bartke A, Yun JS, Wagner TE (1994) Expression of human, but not bovine growth hormone genes promotes development of mammary tumors in transgenic mice. Transgenics 1:431–437

Chandrashekar V, Bartke A (1993) Induction of endogenous insulin-like growth factor-I secretion alters the hypothalamic-pituitary-testicular function in growth hormone-deficient adult dwarf mice. Biol Reprod 48:544–551

Chen HW, Meier H, Heiniger H-J, Huebner RJ (1972) Tumorigenesis in strain DW/J mice and induction by prolactin of the group-specific antigen of endogenous C-type RNA tumor virus. J Natl Cancer Inst 49:1145–1154

Corpas E, Harman SM, Blackman MR (1993) Human growth hormone and human aging. Endocr Rev 14:20–39

Cutler RG (1985) Antioxidants + longevity of mammalian species. In: Woodhead AD, Blackett AD, Hollaender A (eds) Molecular biology of aging. Plenum Press, New York, pp 15–74

DeBoer H, Blok G-J, Van der Veen EA (1995) Clinical aspects of growth hormone deficiency in adults. Endocr Rev 16:63–86

Denckla WD (1974) Role of the pituitary and thyroid glands in the decline of minimal O_2 consumption with age. J Clin Invest 53:572–581

Doherty PC, Bartke A, Dalterio S, Shuster L, Roberson C (1980) Effects of growth hormone and thyroxine on pituitary and testicular function in two types of hereditary dwarf mice. J Exp Zool 214:53–59

Dominici FP, Balbis A, Bartke A, Turyn D (1998) Role of hyperinsulinemia on hepatic insulin binding and insulin receptor autophosphorylation in the presence of high growth hormone (GH) levels in transgenic mice expressing GH gene. J Endocrinol 159:15–25

Drori D, Folman Y (1976) Environmental effects on longevity in the male rat: exercise, mating, castration and restricted feeding. Exp Gerontol 11:25–32

Duffy PH, Feuers RJ, Leakey JA, Nakamura KD, Turturro A, Hart RW (1989) Effect of chronic caloric restriction on physiological variables related to energy metabolism in the male Fischer 344 rat. Mech Ageing Dev 48:117–133

Duffy PH, Feuers RJ, Hart RW (1990) Effect of chronic caloric restriction on the circadian regulation of physiological and behavioral variables in old male B6C3F1 mice. Chronobiol Int 7:291–303

Duquesnoy RJ (1972) Immunodeficiency of the thymus-dependent system of the Ames dwarf mouse. J Immunol 108:1578–1590

Duquesnoy RJ, Kalpaktsoglou PK, Good RA (1970) Immunological studies of the Snell-Bagg dwarf mouse. Proc Soc Exp Biol Med 133:201–206

Esquifino AI, Villanúa MA, Szary A, Yau J, Bartke A (1991) Ectopic pituitary transplants restore immunocompetence in Ames dwarf mice. Acta Endocrinol 125:67–72

Everitt AV, Seedsman NJ, Jones F (1980) The effects of hypophysectomy and continuous food restriction, begun at ages 70 and 400 days, on collagen aging, prolactinemia, incidence of pathology and longevity in the male rat. Mech Ageing Dev 12:161–172

Fabris N, Pierpaoli W, Sorkin E (1972) Lymphocytes, hormones, and ageing. Nature 240: 557–559

Flück C, Deladoey J, Rutishauser K, Eblé A, Marti U, Wu W, Mullis PE (1998) Phenotypic variability in familial combined pituitary hormone deficiency caused by a PROP1 gene mutation resulting in the substitution of Arg- > Cys at codon 120 (R120C). J Clin Endocrinol Metab 83:3727–3734

Grüneberg H (1952) The genetics of the mouse. Martinus Nijhoff, The Hague, pp 122–129

Hayflick L (1998) How and why we age. Exp Gerontol 33:639–653

Holder AT, Wallis M, Biggs P, Preece MA (1980) Effects of growth hormone, prolactin and thyroxine on body weight, somatomedin-like activity and in-vivo subphation of cartilage in hypopituitary dwarf mice. J Endocrinol 85:35–47

Holloway JM, Szeto DP, Scully KM, Glass CK, Rosenfeld MG (1995) Pit-1 binding to specific DNA sites as a monomer or dimer determines gene-specific use of a tyrosine-dependent synergy domain. Genes Dev 9:1992–2006

Hunter WS, Croson WB, Bartke A, Gentry MV, Meliska CJ (1999) Low body temperature in long-lived Ames dwarf mice at rest and during stress. Physiol Behav (In press)

Kimura KD, Tissenbaum HA, Liu Y, Ruvkun G (1997) daf-2, an insulin receptor-like gene that regulates longevity and diapause in Caenorhabditis elegans. Science 277:942–946

Krzisnik C, Kolacio Z, Battelino T, Brown M, Parks JS, Laron Z (1999) The "Little People" of the island of Krk – revisited. Etiology of hypopituitarism revealed. Endocr Genet 1:9–19

Lane MA, Baer DJ, Rumpler WV, Weindruch R, Ingram DK, Tilmont EM, Cutler RG, Roth GS (1996) Calorie restriction lowers body temperature in rhesus monkeys, consistent with a postulated anti-aging mechanism in rodents. Proc Nat Acad Sci USA 93:4159–4164

Lewinski A, Bartke A, Kovacs K, Richardson L, Smith NKR (1984) Further evidence of inactivity of hypothalamo-pituitary-thyroid axis in Snell dwarf mice. Anat Rec 210:617–627

Li S, Crenshaw BE III, Rawson EJ, Simmons DM, Swanson LW, Rosenfeld MG (1990) Dwarf locus mutants lacking three pituitary cell types result from mutations in the POU-domain gene pit-1. Nature 347:528–533

Masoro EJ (1995) Dietary restriction. Exp Gerontol 30:291–298

Meites J (1993) Anti-ageing interventions and their neuroendocrine aspects in mammals. In: Falvo R, Bartke A, Giacobini E, Thorne A (eds) Neurobiology and neuroendocrinology of ageing. J Reprod Fertil Suppl 46:1–9

Micozzi MS (1993) Functional consequences from varying patterns of growth and maturation during adolescence. Horm Res 39:49–58

Miller DB, Bartke A, O'Callaghan JP (1995) Increased glial fibrillary acidic protein (GFAP) levels in the brains of transgenic mice expressing the bovine growth hormone (bGH) gene. Exp Gerontol 30:383–400

Morgan WW, Bartke A, Pfeil K (1981) Deficiency of dopamine in the median eminence of Snell dwarf mice. Endocrinology 109:2069–2075

Naar EM, Bartke A, Majumdar SS, Buonomo FC, Yun JS, Wagner TE (1991) Fertility of transgenic female mice expressing bovine growth hormone or human growth hormone variant genes. Biol Reprod 45:178–187

Ooka H, Shinkai T (1986) Effects of chronic hyperthyroidism on the lifespan of the rat. Mech Ageing Dev 33:275–282

Ooka H, Fujita S, Yoshimoto E (1983) Pituitary-thyroid activity and longevity in neonatally thyroxine-treated rats. Mech Ageing Dev 22:113–120

Orme SM, McNally RJQ, Cartwright RA, Belchetz PE (1998) Mortality and cancer incidence in acromegaly: a retrospective cohort study. J Clin Endocrinol Metab 83:2730–2734

Papadakis MA, Grady D, Black D, Tierney MJ, Gooding GA, Schambelan M, Grunfeld C (1996) Growth hormone replacement in healthy older men improves body composition but not functional ability. Ann Int Med 124:708–716

Patronek GJ, Waters DJ, Glickman LT (1997) Comparative longevity of pet dogs and humans: implications for gerontology research. J Gerontol 52A:B171–B178

Pell JM, Bates PC (1992) Differential actions of growth hormone and insulin-like growth factor-I on tissue protein metabolism in dwarf mice. Endocrinology 130:1942–1950

Pelletier M, Montplaisir S, Dardennell M, Bach JF (1976) Thymic hormone activity and spontaneous autoimmunity in dwarf mice and littermates. Immunology 30:783–788

Pendergast WR, Li Y, Jiang D, Wolf NS (1993) Decrease in cellular replicative potential in "giant" mice transfected with the bovine growth hormone gene correlates to shortened life span. J Cell Physiol 156:96–103

Pfäffle R, Kim C, Otten B, Wit J-M, Eiholzer U, Heimann G, Parks J (1996) Pit-1: clinical aspects. Horm Res 45 (Suppl 1):25–28

Pfeifer M, Verhovec R, Zizek B, Prezelj J, Poredos P, Clayton RN (1999) Growth hormone (GH) treatment reverses early atherosclerotic changes in GH-deficient adults. J Clin Endocrinol Metab 84:453–457

Phelps CJ (1994) Pituitary hormones as neurotrophic signals: anamalous hypophysiotrophic neuron differentiation in hypopituitary dwarf mice. Proc Natl Acad Sci 206:6–23

Promislow DE (1993) On size and survival: progress and pitfalls in the allometry of life span. J Gerontol 48:B115–B123

Reiser KM (1998) Nonenzymatic glycation of collagen in aging and diabetes. Soc Exp Biol Med Minireview, pp 23–34

Rennels EG, Anigstein DM, Anigstein L (1965) A cumulative study of the growth of sarcoma 180 in anterior pituitary dwarf mice. Tex Rep Biol Med 23:776–781

Roberts RC (1961) The lifetime growth and reproduction of selected strains of mice. Heredity 16:369–381

Rollo CD, Carlson J, Sawada M (1996) Accelerated aging of giant transgenic mice is associated with elevated free radical processes. Can J Zool 74:606–620

Romero MI, Phelps CJ (1993) Prolactin replacement during development prevents the dopaminergic deficit in hypothalamic arcuate nucleus in prolactin-deficient Ames dwarf mice. Endocrinology 133:1860–1870

Rudman D, Feller AG, Nagraj HS, Gergans GA, Lalitha PY, Goldberg AF, Schlenker RA, Cohn L, Rudman IW, Mattson DE (1990) Effects of human growth hormone in men over 60 years old. N Engl J Med 323:1–6

Sacca L, Cittadini A, Fazio S (1994) Growth hormone and the heart. Endocr Rev 15:555–573

Samaras TT, Storms LH (1992) Impact of height and weight on life span. Bull WHO 70:259–267

Schaible R, Gowen JW (1961) A new dwarf mouse. Genetics 46:896

Schaufele F (1994) Regulation of the growth hormone and prolactin genes. In: Imura H (ed) The pituitary gland. Raven Press, New York, pp 91–116

Schneider GB (1976) Immunological competence in Snell-Bagg pituitary dwarf mice: response to the contact-sensitizing agent oxazolone. Am J Anat 145:371–394

Schonholz DH, Osborn CM (1949) Temperature studies in dwarf mice. Anat Rec 105:605

Shire JGM (1973) Growth hormone and premature ageing. Nature 245:215–216

Snell GD (1929) Dwarf, a new mendelian recessive character of the house mouse. Proc Natl Acad Sci USA 15:733–734

Soares MJ, Bartke A, Colosi P, Talamantes F (1984) Identification of a placental lactogen in pregnant Snell and Ames dwarf mice. Proc Soc Exp Biol Med 175:106–108

Sonntag WE, Hylka VW, Meites J (1985) Growth hormone restores protein synthesis in skeletal muscle of old male rats. J Gerontol 40:689–694

Sonntag WE, Xu X, Ingram RL, C'Costa A (1995) Moderate caloric restriction alters the subcellular distribution of somatostatin mRNA and increases growth hormone pulse amplitude in aged animals. Neuroendocrinology 61:601–608

Sonntag WE, Lynch CD, Cooney PT, Hutchins PM (1997) Decreases in cerebral microvasculature with age are associated with the decline in growth hormone and insulin-like growth factor 1. Endocrinology 138:3515–3520

Sornson MW, Wu W, Dasen JS, Flynn SE, Norman DJ, O'Connell SM, Gukovsky I, Carriére C, Ryan AK, Miller AP, Zuo L, Gleiberman AS, Anderson B, Beamer WG, Rosenfeld MG (1996) Pituitary lineage determination by the prophet of pit-1 homeodomain factor defective in Ames dwarfism. Nature 384:327–333

Steger RW, Bartke A, Cecim M (1993) Premature ageing in transgenic mice expressing growth hormone genes. J Reprod Fertil Suppl 46:61–75

Strobl JS, Thomas MJ (1994) Human growth hormone. Pharm Rev 46:1–34

Tatsumi K-I, Miyai K, Notomi T, Kaibe K, Amino N, Mizuno Y, Kohmo H (1992) Cretinism with combined hormone deficiency caused by a mutation on the pit-I gene. Nat Genet 1:56–58

Thompson JL, Butterfield GE, Marcus R, Hintz RL, Van Loan M, Ghiron L, Hoffman AR (1995) The effects of recombinant human insulin-like growth factor-I and growth hormone on body composition in elderly women. J Clin Endocrinol Metab 80:1845–1852

Toogood AA, Shalet SM (1999) Growth hormone replacement therapy in the elderly with hypothalamic-pituitary disease: a dose-finding study. J Clin Endocrinol Metab 84:B1–B6

Ultsch M, deVos AM (1993) Crystals of human growth hormone-receptor complexes. Extracellular domains of the growth hormone and prolactin receptors and a hormone mutant designed to prevent dimerization. J Mol Biol 231:1133–1136

van Buul-Offers S, Veda I, van den Brande JL (1986) Biosynthetic somatomedin C (SM-C/IGF-I) increases the length and weight of Snell dwarf mice. Pediatr Res 20:825–827

Viola-Magni M (1965) Cell number deficiencies in the nervous system of dwarf mice. Anat Rec 153:325–333

Webb SM, Lewinski AK, Steger RW, Reiter RJ, Bartke A (1985) Deficiency of immunoreactive somatostatin in the median eminence of Snell dwarf mice. Life Sci 36:1239–1245

Weindruch R, Sohal RS (1997) Caloric intake and aging. N Engl J Med 337:986–994

Werner H, LeRoith D (1996) The role of the insulin-like growth factor system in human cancer. Adv Cancer Res 68:183–223

Winick M, Grant P (1968) Cellular growth in the organs of the hypopituitary dwarf mouse. Endocrinology 83:544–547

Wolf E, Kahnt E, Ehrlein J, Hermanns W, Brem G, Wanke R (1993) Effects of long-term elevated serum levels of growth hormone on life expectancy of mice: lessons from transgenic animal models. Mech Ageing Dev 68:71–87

Yamashita T, Nifuji A, Furuya K, Nabeshima Y, Noda M (1998) Elongation of the epiphyseal trabecular bone in transgenic mice carrying a klotho gene locus mutation that leads to a syndrome resembling aging. J Endocrinol 159:1–8

Ye L, Nakura J, Morishima A, Miki T (1998) Transcriptional activation by the Werner syndrome gene product in yeast. Exp Gerontol 33:805–812

Stem Cells and Genetics in the Study of Development, Aging, and Longevity

Gary Van Zant

1
Introduction

1.1
Definitions

Development and aging, indispensable in the case of the former and re-grettably unavoidable in that of the latter, are inextricably linked. Most fundamental to this relationship is that both are strictly defined by the same temporal dimension. Development in mammals is generally regarded as be-ginning at the union of egg and sperm producing a zygote, and ending at or near the time of birth; thus, this process occurs largely in the intrauterine environment of the mother. Some may argue that this definition is too re-strictive and that development continues until full body size is attained or, perhaps more importantly, until reproductive function commences. Aging is less precisely defined both conceptually and temporally. It is important to be precise conceptually about the meaning of aging. It is not the same as or-ganismal longevity nor the determination of longevity. The latter, for ex-ample, has a strong genetic component that is usually considered to be lacking in aging, a process largely defined by the effects of environmental factors. There are congenital diseases, the progerias, in which aging is greatly accelerated, but the defect is typically in a mechanism repairing damage from environmental stresses (Martin et al. 1996). Some regard aging as the im-mediate events that lead up to an organism's death, including the diseases of later life that usually claim an organism; in humans these are often cardio-vascular disease, cancer, complications from diabetes, and complications arising from neurodegenerative diseases and dementia. For the purposes of discussion here, however, a broader interpretation will be applied and it will be assumed that aging begins at conception, since the deleterious side effects of forging an existence probably begin to accumulate immediately. Thus,

Blood and Marrow Transplant Program, Lucille P. Markey Cancer Center, University of Kentucky Medical Center, 800 Rose St., Lexington, KY 40536-0093, USA

Results and Problems in Cell Differentiation, Vol. 29
Hekimi (Ed.): The Molecular Genetics of Aging
© Springer-Verlag Berlin Heidelberg 2000

according to these admittedly arbitrary definitions, development and aging occur simultaneously during the first part of a mammal's life; aging not only persists but probably accelerates.

In support of this interpretation, several chapters in this Volume reflect on the cumulative, deleterious effects of the toxic by-products of oxidative metabolism and environmental stress on an organism's longevity. Life-sustaining metabolism, of course, is already active at conception and occurs throughout life, although damage in the form of oxidized biochemical cell components does not necessarily accumulate at the same rate throughout life. Efficient mechanisms have evolved to protect individual cells, and ultimately multicellular organisms, from reactive oxygen species and other environmental stresses. It is the erosion of these defenses that may lead to accelerating damage later in life. Indeed, as has been noted in recent reviews on the subject, decline in potential longevity is the net balance between damage and the ability to repair it (Hayflick 1998; Johnson et al. 1999). It is for precisely this reason that a tight focus on diseases of the elderly may not address the fundamental changes in physiology that loosen an organism's defenses to disease.

1.2
Cancer as a Disease of Both Development and Aging

Cancer may be an exception here. It is typically a disease of both the old and, to a lesser extent, the young, but not particularly those in their middle years, and may exemplify the vulnerability of the young and those of advanced age to cellular dysfunction, but through different mechanisms. As discussed in detail in another chapter (9; Dolle et al.), cancer is a cellular disease usually arising in one or at most a few cells of a dividing tissue which have received damage that causes genomic instability. The robust proliferation in many tissues of children exposes them to an increased risk of genetic damage; the old may be vulnerable due to diminished ability to repair genetic damage. Thus, cancer exploits both ends of the damage/repair spectrum where extensive growth in the young presents a large target for genetic damage despite defenses that are intact; compromised defenses may do in the old despite a general decline in proliferation of cells in many tissues (Rubin 1997; Fig. 1).

In further support of a broad definition of aging is the notion that a mitotic clock (or clocks) governs cellular senescence in vivo and, by extension, is somehow related to organismal longevity. The use of the word clock is unfortunately misleading, since it implies that passage of time is the only relevant issue, whereas the cumulative number of mitoses, and thus replicative stress on the tissue, may be more important (Hayflick 1998). This notion will be further developed later in this chapter. The bulk of mitotic activity in many organ systems is thus temporally concentrated during

Fig. 1. Cancer risk as a function of time. Development and aging contribute risk factors early and late in life to affect the chance of genetic instability and cancer

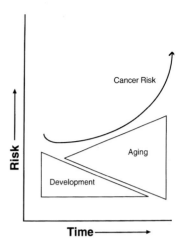

development. The brain has classically served as an example here since it attains nearly full size relatively early in development and cells comprising the organ have been thought of as postmitotic and not capable of substantial further expansion or regeneration. At the other end of this hypothetical spectrum, the skin, the gut lining, and the lympho-hematopoietic system have been regarded as hallmark organs, since they are in a continuous developmental state even in the adult. Such intense proliferative activity is necessitated by the short life spans of the individual mature cells in these tissues, and by their large numbers. As a point of reference, nearly 1×10^{12} red and white blood cells are lost and must be replaced daily in a healthy human (Goldwasser 1975; Metcalf 1988).

1.3
Stem Cells Are Life-Sustaining Vestiges of Organismal Development

This concise definition of tissues as being either nonrenewing (the majority) or renewing (the minority) has recently been challenged by two growing bodies of evidence. First, stem cells have been isolated and characterized from a lengthening list of organ systems, the central nervous system and the liver to name just two (Johansson et al. 1999). Second, it appears that stem cells possess remarkable plasticity in the developmental programs they are capable of carrying out; moreover, program selection is largely regulated by the environment in which they find themselves. For example, a recent report demonstrates that at least in some mice, stem cells derived from the central nervous system, when injected into irradiated recipients, are capable of producing the cornucopia of mature blood cells normally produced by hematopoietic stem cells (Bjornson et al. 1999). The converse, production of neural cells by hematopoietic stem cells, has also been reported. Similarly, transplanted hematopoietic stem cells from adult bone marrow have been shown to participate in rat liver regeneration after partial hepatectomy using carbon tretrachloride (Peterson et al. 1999). In this study, hepatic oval cells, progenitors for hepatocytes and bile duct cells, were shown to be derived from intravenously injected marrow cells. Thus a growing number of adult tissues exhibit characteristics usually associated with nascent tissues during development.

1.4
Interrelatedness of Development and Aging – Chapter Outline

If aging begins at conception, can development not continue until death? In the following sections, I wish to develop this theme further. I wish to first define and update the concept of development as a process of sequential and ordered restriction in developmental potential. Next I wish to define the stem cell population as the pivotal locus in a developmental system, whether it be a whole organism undergoing embryogenesis or an adult, self-renewing tissue. I wish to review the evidence identifying the stem cell population as a critical focus for the age-related accumulation of damage due to reactive oxygen species (ROS), environmental and replicative stress and subsequent genetic damage and instability. I wish to discuss the relationship between proliferation of stem cells and cellular senescence and provide evidence that the stem cell population ages. Lastly I wish to discuss our results and those of others studying stem cells of the blood-forming hematopoietic system as a model system, particularly the implications this population may carry for organismal longevity and aging patterns in the mouse.

2
Development as a Reversible Restriction of Developmental Potential

2.1
What Is the Mechanism?

The conundrum of developmental biology ever since DNA was first identified as the genetic material in the 1950s was, and continues to be, how developmental programs restrict and selectively direct developmental potential. It may be advantageous to take a similar conceptual approach to at least some aspects of the study of aging and place an emphasis on developmental processes, as has been suggested (Schlessinger and Ko 1998). This approach has been successful in *Caenorhabditis elegans*, for example, as illustrated by the finding that longevity could be increased 60% by laser ablation of the two germ line cell founders, but leaving intact the founders of the somatic gonad (Hsin and Kenyon 1999). The authors hypothesize that the gonad proper regulates longevity systemically through secretion of a factor acting in the DAF gene pathway.

An early idea explaining restriction of developmental potential was that genetic material was sequentially lost during development and that a cell's potential was restricted by the lost genes (Weismann 1893). However, when Gurdon first showed that the nucleus from a gut cell of a tadpole could, when implanted into an enucleated egg, give rise to a complete and apparently healthy, adult frog, it became clear that genetic material was not physically lost (Gurdon 1968). Rather, with rare exceptions, all cells in an organism contain the same repertoire of genes as was present in the zygote at conception. Exceptions include the red blood cell which lacks a nucleus entirely, and clonally selected lymphocytes which have undergone genetic recombination restricting antibody repertoire. The processes regulating development are now understood to involve the selective activation of some genes and the silencing of others principally through complex sets of transcriptional activators and repressors (Mannervik et al. 1999). It is evident from the original frog experiment and from more recent experiments in sheep and mice that cloning an individual organism by drawing upon the complete genetic heritage of even differentiated cells is possible through the reversible nature of commitment to a given developmental lineage (Wilmut et al. 1997; Wakayama et al. 1998; Wakayama and Yanagimachi 1999). In other words, the intracellular microenvironment, if carefully chosen, is capable of directing which genes are and which are not transcribed. Even more remarkably in the case of animal cloning, this dynamic process is directed such that it follows the precise time schedule necessary to produce a new individual. Such experiments also provide an important framework in which to study the effects of aging on the genome. If nuclei from adult cells can be re-programmed to recapitulate the entire developmental program of a species, are biological markers of aging, such as telomere length, reset to embryologic settings in the process? Or, for example, do mice or sheep cloned from such nuclei have a deficit in life span corresponding to the age of the donor animal? Emerging results from monitoring age-related changes in sheep cloned from adult epithelial cell nuclei suggest the latter (Shiels et al. 1999). Three years after the birth of Dolly and her cohorts, shortened telomeres in somatic cells of these animals reflect the age of the nuclei donor plus their chronological ages. Dolly, for example, is 3 years old, but has telomeres shortened to the length found in 9-year-old sheep. Thus, it appears from this admittedly small sample that telomere lengths in an adult nucleus are not reset when transplanted into the embryonic environment of a developmentally early cell. If substantiated, these findings make it likely that cloned animals will undergo premature aging and that their life spans may be foreshortened.

2.2
Developmental Choices Are Not Necessarily Immutable

The existence of identical twins is formal proof that full developmental potential is naturally retained through at least the first several divisions of the
zygote, in the absence of any experimental manipulation. In mice, this has
been taken advantage of in the laboratory to derive embryonic stem (ES) cell
lines from preimplantation-stage embryos called blastocysts, or from primordial germ cells (PGCs), those cells destined to differentiate into sperm or
oocytes. Such cells have been invaluable in the advances that have been made
in genetic studies involving transgenic, knockout, and knockin mice wherein
selected genes are overexpressed, silenced, or substituted, respectively. It is
now known that even cells from adult tissues can, when injected into
blastocysts, not only survive but contribute to their line of differentiation at
the appropriate time in development. Geiger et al. (1998) showed that adult
hematopoietic stem cells, in a blastocyst environment, were reprogrammed to
participate in embryonic and fetal blood cell formation by generating progeny synthesizing embryonic and fetal hemoglobins at the appropriate times
in development. Conversely, stem cells from embryonic tissues gave rise to
progeny which synthesized adult hemoglobin rather than embryonic or fetal
hemoglobin.

Recent experiments involving ES cells and PGCs from humans show that
cloning of individuals is at least scientifically possible (Thomson et al. 1998;
Shamblott et al. 1998), if not socially acceptable. Of more immediate and
practical value, however, these studies demonstrate the potential for medical intervention in the treatment of a number of diseases involving acquired or congenital genetic damage in virtually any tissue. By combining
the knowledge gleaned from frogs, mice, sheep, and more recently, humans,
it may soon be possible to produce custom-made cells to replace or repair
a defective tissue in patients (Rossant and Nagy 1999; Solter and Gearhart
1999). One possible scenario, not without formidable ethical and legal
dilemmas, would be to first obtain a nucleus from a differentiated somatic
cell from virtually any tissue of a patient (e.g., an epithelial cell or a
fibroblast), transplant it into an enucleated human ovum (not necessarily
from a related donor) and in vitro develop an embryo to the blastocyst
stage, at which time a custom ES cell line could be derived. Since ES lines
are pluripotent, differentiation could be directed in vitro to the appropriate
tissue according to the patient's needs; for example, pancreatic islet cells to
treat diabetes. According to specific need, as with mouse ES cells, the
genomes of human ES cells could be manipulated in vitro to replace a
congenitally defective gene or overexpress a specified gene. Such cellular
therapeutics would have the distinct advantage over current transplantation
therapies of using a syngeneic graft, thus obviating any graft-host histo-
incompatibilities.

3
Stem Cell Populations Drive Developmental Systems

3.1
Models of Stem Cell Differentiation

In adult, self-renewing tissues, the stem cell population serves as a continuous source of the appropriate differentiated cells throughout the life span of the animal. Thus, over a generally long period of time, this small cell population, acting in response to a complex series of regulatory controls both cell-intrinsic and cell-extrinsic, must parcel differentiation and proliferation in such a way as to ensure its survival while at the same time providing enough mature cells. It is not the aim here to provide an exhaustive review of stem cell models; for this the reader is directed to a volume of reviews in which these issues are more fully developed (Potten 1997). However, in brief, a stem cell population may meet these demands through two broad mechanisms:

3.1.1
Clonal Succession

One specifies that the majority of stem cells are maintained in a quiescent, noncycling state until needed and then one (or a few) at a time is recruited and activated to become the active clone(s) (Kay 1965). Such a clonal succession model preserves primitive developmental potential in the founder stem cell population through dormancy, but without additional features does not allow for expansion of the stem cell pool through self-renewal. It also implies that, despite extensive developmental potential of individual stem cells and their resulting progeny, clonal longevity is limited and individual clones thus have a finite life span.

3.1.2
Flexibility in Types of Daughter Cells Produced by Stem Cell Division

The second mechanism relies on stem cell proliferation to provide flexibility and preserve a stem cell reserve. At each such cell division, depending upon demands, the stem cell, through symmetric or asymmetric divisions, may give rise to (1) new stem cells, (2) progenitor cells one step removed from stem cells but potentially still pluripotent, or (3) a combination of the two cell types. Despite the fact that Osgood proposed such a model of alpha cell division in the 1950s (Osgood 1957), little is known of how the regulatory controls might work. Signaling mechanisms, if any, specifying which type of cell division a stem cell is to carry out are presently obscure. It could be a stochastic event or it could be deterministic in nature through responses to extrinsic signals derived from the local milieu (cytokines and stroma). Alternatively, deterministic signals could be generated intrinsically by the stem cell itself; for example by partitioning of a cell component at mitosis. Such a mechanism is employed at the earliest cell divisions of embryogenesis when preformed cell components of the zygote are partitioned to specify the earliest restrictions in cell fate among blastomeres (Beddington and Robertson 1999). The significant role that intrinsic telomere dynamics may have on this process are only now being appreciated. The role of telomeres in aging and longevity are discussed in more detail below.

3.1.3
Stem Cell Populations Reflect Physiological Need While Developmental Choices at the Individual Cell Level May Be Stochastic

When the stem cell population is viewed as a whole, a coordinated response to specific hematopoietic needs may be concerted and highly effective, whereas at the level of individual stem cells all mitotic options, symmetric and asymmetric, may be open to each cell but with variable probabilities that reflect the current needs (Till et al. 1964). It should be pointed out that at least the major features of these models do not necessarily need to be mutually exclusive. For instance, the active stem cell clone in a succession model may have the options open to it presented in the second mechanism, provided that stem cell division may produce other stem cells, an admittedly significant departure from the tenet. Irrespective of the single or multiple mechanism(s) of stem cell control, we know from a large body of experimental and clinical data that stem cell-driven developmental systems have considerable inherent flexibility. In fact, it appears to be sufficiently extensive that maintenance of the stem cell population should be assured under a variety of experimentally created and naturally occurring conditions including the current topic, organismal aging.

3.2
Why Are Stem Cells Difficult to Study?

Because of their numerical rarity in tissues and their lack of distinguishing morphological characteristics, stem cells have been, and continue to be, difficult to study. As has been previously noted, because of these mitigating properties, stem cells have been studied by their functional abilities – proliferative and developmental potencies. Unfortunately, such measurements harken back to Heissenberg's uncertainty principle in physics, which states that the act of taking a measurement of something alters its properties. It is certainly true that an assay of stem cells that requires their differentiation, precludes the use of a given population from other use. Historically stemming from concerns about the health effects of radiation following the advent of nuclear energy and the use of nuclear weapons, the hematopoietic stem cell has been the focus of concerted studies over the years and has arguably yielded the largest body of information about stem cells in general. For this reason and because of a personal interest in hematopoietic stem cells, this chapter concentrates on this stem cell type. However, as already alluded to, revelations of extensive flexibility in developmental potential make it increasingly likely that findings obtained with one stem cell will be generally applicable to others.

Despite considerable efforts, it has only been in the last 10 years that stem cell purification has been widely available to workers in the field, permitting their detailed prospective study. Expression of one set of cell surface antigens, and the lack of expression of other differentiation markers, was correlated with functional stem cell activity such that it became possible to use fluorochrome-tagged antibodies and fluorescence-activated cell sorting and/or immuno-absorptive methods to isolate viable stem cells. Due to the pioneering developmental efforts of several labs, it is now reasonably straightforward to obtain purified stem cell populations (Visser et al. 1984; Spangrude et al. 1988; Ploemacher and Brons 1988). They contain virtually all of the long-term repopulating ability of bone marrow as measured by transplantation into lethally irradiated mice (Uchida and Weissman 1992), or, in the case of human stem cells, to durably engraft immunodeficient mice bearing mutations for nonobese diabetes and severe combined immunodeficiency (NOD-SCID mice; Dick et al. 1995). Similar advances have recently been made in the purification of neural stem cells, drawing upon the techniques used for hematopoietic stem cell isolation (Morrison et al. 1999).

3.3
Renewal of Stem Cells as Revealed by Transplantation

Is it really true that maintenance of a viable stem cell population is assured under all manner of normal and not-so-normal life situations? A widely accepted assumption held by most stem cell biologists is that stem cells are generally capable of self-renewal such that daughter stem cells (but not derived progenitor cells) are functionally equivalent to those of the previous generation. There is good reason for this; support for the concept, in part, comes from experience with both mice and humans in the field of bone marrow transplantation. In this procedure, a miniscule fraction of the original stem cell pool is transplanted and is required to satisfy not only the immediate need for mature blood cells in the ablated recipient, but also to rebuild the coordinated hierarchy of stem, progenitor, and maturing cells necessary to maintain lympho-hematopoiesis over the long term. That such a small stem cell graft can accomplish this formidable task appears to beg the question as to whether stem cell renewal occurs. Limiting dilution transplant experiments using highly purified stem cells have convincingly shown that even a single cell is capable of this dramatic feat (Spangrude et al. 1995; Osawa et al. 1996). Moreover, serial transplant studies in mice have shown that the original graft may not only repopulate the primary host but if marrow from these recipients is subsequently harvested and transplanted into and passed through secondary, tertiary, and quaternary hosts, the original, small stem cell population can cumulatively provide differentiated cells for longer than the life span of the original donor animal (Siminovitch et al. 1964; Harrison 1972).

Consequently, there is little question that a functional stem cell population can increase in size and therefore that stem cells can replicate to produce other cells with the cardinal functional characteristics of stem cells. Considerable light has been shed on this issue by the use of stem cells that have been genetically marked by proviral insertion. Since the genomic insertion pattern for each stem cell and its progeny is unique, stem cell pedigrees can be established. Szilvassy et al. (1989), Fraser et al. (1990), and Smith et al. (1991) have all shown that retrovirally marked stem cells could replicate either in vitro or in vivo and that their progeny retained developmental potency to subsequently engraft transplant recipients long-term. However, there is also abundant data showing that the stem cell population is qualitatively and quantitatively diminished after transplant (Spangrude et al. 1995). The fact that one cannot serially transplant stem cells indefinitely is but one finding supporting this. Long-term repopulating stem cells recover after transplant to only about 4% of normal values in 6 weeks, and do not further increase even after many months (Iscove and Nawa 1997). Despite the blunted recovery, absolute numbers do increase by about ten-fold over the numbers in the graft used for each of the transplants. The fact that the ten-fold expansion is found irrespective of the number of stem cells in the graft, led Iscove and Nawa to propose that extrinsic factors in the recipient, such as cytokines, limited the level of expansion rather than replicative stress, exhaustion, and senescence.

Retrovirally marked stem cells have also been used to quantify contributions of individual stem cells to engraftment after transplant. Taken in sum, the results have showed that short-term engraftment was accomplished by many individual clones but that after a few months an oligoclonal pattern emerged (Jordan and Lemischka 1990). Some, but not all, clones persisted for periods of time that approximated the mouse life span (Capel et al. 1989, 1990), and in extreme examples a single clone maintained all hematopoiesis in a transplant recipient for a year or more (Keller and Snodgrass 1990).

3.4
Stem Cell Kinetics in Steady-State Animals May Be Different

Before accepting data from transplant models as reflecting the normal steady state, it must be acknowledged that fundamental differences may exist between the two settings which may affect stem cell regulation. The disparity in the sizes of the natural stem cell pool compared with the small number in a graft has already been alluded to. Moreover, it has been argued that the experimental manipulations involved in carrying out a stem cell transplant, with respect to both the donor and the recipient, are sufficient to make extrapolation to natural controls invalid (Harrison et al. 1978). The disruption of the stem cells from their close association with stromal cells comprising their local microenvironment, their injection into the venous circulation of the recipient, and the perhaps selective requirement that stem cells lodge in an appropriate new microenvironmental niche, introduces large variation from steady-state conditions. Similarly for the recipient, treatment with large doses of ionizing radiation and/or chemotherapeutic drugs has significant but not fully defined effects on the stromal cells and the resulting cytokine titers to which the stem cells are exposed.

In contrast to results obtained from serial transplantation, Ross et al. (1982) showed that a protocol involving 25 cycles of hydroxyurea injections over a year did not measurably diminish stem cell function, despite the fact that the cumulative replicative requirement was quantitatively similar to serial transplantation. These data suggest an inherent difference between experimental measurements of stem cell function made using transplantation models and those made in vivo without transplantation.

Consistent with these results were findings obtained with chimeric mice constructed by combining two ES cell lines bearing distinguishable markers expressed in mature blood cells, but otherwise sharing the same genetic background (Harrison et al. 1987). They were characterized by stable, chimeric blood cell populations over long periods of time, suggesting the simultaneous contributions of a large number of stem cells of both genotypes rather than oligoclonal participation and clonal succession. Study of chimeras constructed by aggregating embryos of different mouse strains (allophenic mice) have provided data consistent with both models (Warner et al. 1977; Van Zant et al. 1990). Some chimeras, at certain times in their life histories, exhibited dramatic fluctuations in chimerism of blood cells, consistent with a clonal succession model, whereas the majority displayed chimerism consistent with the simultaneous participation of a large number of stem cell clones. As discussed below, the use of allophenic mice has provided insight into competitive advantages of genotype-restricted stem cell clones, since in this model embryos of different genetic backgrounds were combined, and has led to the identification of several stem cell genes affecting their cell cycle kinetics and population sizes in the bone marrow.

3.5
Steady-State Stem Cells Enter and Leave Cell Cycle Regularly

In apparently decisive studies resolving the issue of stem cell cycling, Bradford et al. (1997) and Cheshier et al. (1999) have built on the original findings of Pietrzyk et al. (1985), by labeling hematopoietic stem cell populations of normal, unmanipulated mice through the long-term addition of bromo-deoxyuridine (BrdUrd) to their drinking water. The results have shown that essentially the entire stem cell population is slowly labeled over 2–3 months in a pattern consistent with virtually all cells having passed through at least one cell cycle during this period. Given the fact that only a small percentage of stem cells are in cycle at any given time, the data suggest that stem cells may repetitively enter and leave cell cycle from a quiescent G_0 state. In contrast, rapidly dividing mouse cells usually have a 12–24-h cell cycle time, with essentially none being out of cell cycle in quiescent G_0. These data argue strongly against a succession of stem cell clones, since such a mechanism would show labeling in only a small fraction of stem cells during the period of exposure to BrdUrd – in theory, perhaps only one. The results also corroborate the findings in chimeras showing stable blood cell chimerism over long time periods and further underscore the fundamental difference between stem cell regulation in normal animals and in a transplant model.

3.6
Are Large Animals Fundamentally Different?

Much of the experimental evidence on stem cell usage issue comes from studies of mice, but there is some evidence in larger mammals consistent with an oligoclonal, active stem cell population. For example, autologous transplant studies in cats have shown that allelic expression of an X-linked marker in marrow cells of heterozygous females fluctuates for several years after transplant, but with time becomes stable and mature blood cell populations may express only one or the other G6PD alleles (Abkowitz et al. 1990). These results, like the pattern of clonal expansion of genetically marked and transplanted stem cells in mice, are consistent with few stem cells clones being simultaneously active both short- and long-term after transplant. Such an experimental approach took advantage of the fact that only one X chromosome is transcriptionally active in adult cells and that X-inactivation is random in embryonic cells, including the founder hematopoietic stem cell population. Similarly, in older women heterozygous for allelic variants of G6PD, there is increased and apparently random skewing toward one variant or the other during aging (Gale et al. 1997). Such data are consistent with increasingly fewer clones contributing to hematopoiesis in old age, and provide additional evidence suggesting a small, active stem cell population.

However, it seems unlikely that stem cell population organization and usage would be fundamentally different between mice and larger mammals including the human. Rather, given the different methods of study, technical details of experimental procedures and assays may account for the differences. For example, in the transplantation studies in cats, stem cells were transplanted at limiting dilutions such that some recipients received too few stem cells to engraft necessitating a second transplant with more stem cells. It has been shown previously that the number of stem cells transplanted affects not only the clonal distribution of cells amongst the host's hematopoietic tissues, but also the contributions from individual cells during engraftment (Micklem et al. 1987). As one might expect, the greater the number of stem cells transplanted, the larger the number contributing to engraftment. At limiting dilutions, oligoclonal hematopoiesis might be expected to be the norm as illustrated experimentally in Micklem's elegant study.

The issue of X chromosome inactivation and somatic cell clonality in elderly women may be complicated by additional factors. Recent studies have shown that the X chromosome may contain genes relevant to allele-specific skewing that are deterministic rather than representing stochastic events or a diminution in the number of active clones over time. In accordance with other studies, Buller et al. (1999) found nonrandom X chromosome inactivation patterns usually interpreted as evidence of the clonal origin of tumors. However, they hypothesize that the data in this case are consistent with the existence of an X-linked tumor suppressor gene. A germline mutation of this gene combined with nonrandom X-inactivation may reduce or abolish expression of the wild-type allele and predispose these women to ovarian and/or breast cancer. Underscoring the importance of deterministic X-linked alleles, it has been shown in certain heterozygous cats that clonal dominance during aging is due to a competitive advantage conferred by a locus on the X chromosome that possibly acts by enhancing proliferation or decreasing cell death in the affected clone (Abkowitz et al. 1998).

3.7
What Role for Apoptosis in a Continuously Renewing Stem Cell System?

The foregoing discussion has centered on the issues of stem cell division and the developmental cell fate of stem cell progeny. The role of apoptosis has been neglected although it likely plays an important role in determining population dynamics of stem, progenitor and maturing cell populations, as it does during development (Vaux and Korsmeyer 1999). Conditional survival of cells at all levels in the developmental hierarchy, including stem cells, may involve a variety of cell-extrinsic and cell-intrinsic factors preventing or enabling apoptosis. Dysregulation of tumor-suppressing genes in cancer, allowing tumor cells to bypass normal population growth restraints through apoptosis, demonstrates the importance of cell death in the process of population control. Indeed, the natural limitation in vivo of sufficient erythropoietin, the hormone regulating red cell production, may normally doom large numbers of developing erythrocytes to an apoptotic death under steady-state conditions, but at the same time provide a mechanism for a rapid response to an acute need for red cells mediated by increased erythropoietin production (Koury 1992; Yu et al. 1993).

4
Stem Cell Populations as Critical Targets of Damage During Aging

4.1
Cancer and Congenital Diseases

Since stem cells in any organ are the sole source of renewable, differentiated cells, genetic alterations induced in stem cells have broad implications for the tissue as a whole. Depending on the number of simultaneously active stem cells, genetic changes in one will be inherited by significant segments of the entire mature populations. This may especially be true in recipients of hematopoietic stem cell transplants in which few stem cells, perhaps only one, may contribute to long-term hematopoiesis. If the affected clone is endowed with a proliferative advantage either as a consequence of the induced damage or in the form of a "second hit" such as a mutation in Ras, p53, pRB, or induction of telomerase synthesis, the clone may become the predominant supplier of mature cells. Several of the leukemias and my-elodysplastic diseases, as well as heritable diseases such as sickle cell disease, are the consequence of genetic changes manifested through the stem cell population. Sickle cell disease is due to a congenital globin chain mutation whose debilitating symptoms are the result of red blood cells derived from stem cells possessing the defective gene. Of course, in a congenital disease all cells of an affected individual carry the defective globin gene but it is in the blood-forming tissues that the mutation is manifested. Instances of genetic damage originating in the stem cells themselves include the secondary leukemias that may result from the intensive radio- and chemotherapy cancer patients undergo in treatment of their primary malignancy. Treatments designed to kill tumor cells by interfering with their ability to replicate, unfortunately, also have residual genetic effects on normal cells not killed outright by treatment. It is believed that the secondary leukemias are the result of genetic damage to one or few hematopoietic stem cells and are manifested only many years after the treatments ended (Socie 1996; Anderson et al. 1997).

4.2
Hematologic Malignancies

Studies of the hematologic malignancies – leukemia, lymphoma, myelodys-plastic syndromes in particular – have provided strong evidence that genetic mutation and genomic instability in one stem cell can and may result in malignancy. It is also clear that the barriers to such a neoplastic transforming event are formidable and that its occurrence and persistence as a cell clone is rare. Molecular markers identifying neoplastic cells have convincingly shown that these malignancies are clonal and are the consequence of events occurring in hematopoietic stem cells and their immediate progeny. It is interesting to note that despite a life-long requirement for intestinal epithelial cells that is roughly equivalent to the numerical requirement for blood cells, intestinal tumors are less common than in the hematopoietic system (Loeffler and Potten 1997). Whether or not this is due to more effective surveillance, repair and/or deletion of defective clones, is not known. What is known is that the probability of either type of malignancy occurring increases with age. It is likely that this increased risk is partly due to cumulative damage to numerous cellular components, including the genetic material, caused by a variety of intracellular and environmental factors. Genetic material includes not only that in the nucleus but in the mitochondria as well (Bohr et al. 1998). Damage to the DNA of the latter may not only directly affect metabolism of the affected cell but is afforded a cytosolic means of inheritance. Affected mitochondria may be preferentially parceled to one or both daughter cells at mitosis.

4.3
All Reactive Oxygen Species Are Not Bad

As discussed in chapters 2 (Jazwinski), 3 (Sohal et al.), and 8 (Ishii and Hartman) in this Volume, ROS are responsible for cumulative damage to numerous cell components. However, there is growing evidence that ROS may serve important positive cellular roles as well. For example, Sattler et al. (1999) have shown that several important hematopoietic cytokines rely on intracellular pathways in which ROS is a normal and necessary signal. Generation of ROS is a normal consequence of these cytokine-receptor interactions, particularly in signaling pathways involving tyrosine phos-phorylation, and ROS contribute to downstream signaling events. Such findings point up the complexity of normal cellular mechanisms and argue that indiscriminate elimination of ROS may have deleterious effects as well as antiaging effects.

5
Hematopoietic Stem Cells as a Model Population for Studies of Aging

5.1
Availability and Ease of Study

Hematopoietic stem cells have been the subject of extensive studies, including effects of aging, for several reasons; (1) their mature progeny, blood cells, are readily accessible in the peripheral blood and this source can be sampled repetitively from the same animal or patient over long time periods; (2) stem cell transplantation is used extensively in the clinical treatment of cancer, heightening research in the area; (3) the hematopoietic tissues containing stem cells, principally bone marrow in adults, are reasonably accessible.

The diffuse nature of hematopoietic tissue has been both a curse and a benefit to its study. The diffuse organization has made it difficult to establish the histological organization of the marrow, particularly with respect to the interactions of hematopoietic cells and supporting stroma and vasculature. Compare this with the organization of the crypts and villi of the intestinal epithelium where the locations and numbers of stem cells, progenitors, and maturing cells follow precisely defined and discernible patterns (Potten and Loeffler 1990; Winton and Ponder 1990). On the other hand, hematopoietic tissues are easily handled and, most importantly, it is very easy to obtain viable single cell suspensions. The latter point has allowed the field to benefit from flow cytometry, a powerful analytical and cell separation tool. Combined with advances in cell culture, these have led to an increasingly detailed, but still incomplete, developmental map of the system.

5.2
Contradictory Data

The path to a clear view of the effects of aging on the hematopoietic stem cell population has been, and remains, difficult [for additional perspectives, see Morrison et al. (1996) and Globerson (1999)]. Whether or not stem cell populations age is an important question, with significant clinical ramifications, to which answers are still being sought. There is a wealth of data on this issue, with some of it being contradictory. On one hand, evidence points to an essentially unlimited renewal capacity, while on the other there are clear indications of decremental functional responses to replicative stress. The contradictions arise, at least in part, because of the myriad technical approaches taken and assays used in this pursuit. Diversity rather than a more standardized approach is rooted in the scarcity of stem cells and, at least historically, their lack of easily distinguished characteristics. More rapid progress has come since fairly standardized cell separation protocols have been available for the purification of mouse and human stem cells. As previously discussed, using flow cytometry and sorting, it is now possible to obtain fairly large populations of highly purified stem cells for research and even clinical transplantation.

5.3
Stem Cell Populations Age

Although the size and composition of mature blood cell populations remains largely unchanged into old age, functional attributes of some mature cells is compromised (Williams et al. 1986). For example, biochemical changes in red cell membranes shorten their life span in the circulation, which may alter their oxygen exchange rates and capacities (Clark 1988). In addition, the distribution of lymphocyte subpopulations is altered, the cytokine profiles they produce are altered, and at least some aspects of surveillance and disease fighting are diminished (Miller 1996; Effros 1998).

But what of the stem cells in this regard? Despite their ability to continuously supply mature cells at normal levels throughout life and their ability under extreme experimental conditions such as bone marrow transplantation to meet demands not normally encountered, the literature generally, but not universally, supports the contention that the stem cells age. For example, early studies by Ogden and Micklem (1976) determined the rate of decline in stem cells during serial transplantation of marrow from young and old donors. They found that stem cells in old marrow were capable of fewer repetitive transfers than those in young marrow. Similarly, Micklem et al. (1972) found a preferential engraftment advantage using fetal liver rather than adult mouse marrow to transplant lethally irradiated recipients. In concurrence, Mauch et al. (1982) found similar decrements in the proliferative capacity of marrow from old animals when transplanted.

However, other studies of the functional properties of bone marrow from aged mice, showed that old stem cells engrafted irradiated recipients at least equally as well as, and in some cases more effectively than, marrow from young adult mice (Harrison 1983). Central to these and to later studies helping to clarify the issue, Harrison (1980) developed a competitive repopulation assay in which mixtures of young and old marrow or fetal liver were used to uncover potential functional differences in cell populations to engraft and maintain hematopoiesis long-term. Distinguishable genetic markers between the cell sources enabled one to quantify and determine the genetic derivations of hematopoietic cells.

The original study showing no diminished capacity of old stem cells was carried out by mixing specified numbers of marrow cells from young and old marrow to comprise the graft. Several groups have subsequently shown that the number of stem cells, identified either by cell surface phenotype or functionally, in most mouse strains actually increased with age (Harrison et al. 1989; Morrison, Wandycz 1996; de Haan 1997). Thus, at least part of the repopulation results could be explained by the fact that the frequency of stem cells was higher in aged marrow and therefore had an unaccounted for numerical advantage, irrespective of any qualitative difference that may exist between young and old stem cells.

Additional studies have contributed to a growing consensus pointing to stem cell aging. As further examples, Rebel et al. (1996) compared murine fetal liver and young adult marrow stem cells in a limiting dilution repopulation assay and found an extensive functional advantage of the fetal cells. Similarly, Chen et al. (1999) used modeling techniques based on Poisson probabilities to analyze repopulation by paired admixtures of fetal liver, young bone marrow, or old bone marrow. They found an inverse correlation between donor age and functional repopulating capacity in BALB/cBy mice. Fetal liver stem cells had up to three times the functional capacity of young marrow stem cells, which, in turn, had twice the capacity of stem cells from marrow of 2-year-old animals. Morrison et al. (1996) showed that despite as much as a five-fold increase in stem cells in old marrow, the sorted cells had only about one-fourth the engrafting potential when transplanted.

Lansdorp's group has sorted human and mouse stem cells to high purity from hematopoietic tissue of different chronologic age including fetal liver, umbilical cord blood, and bone marrow. Candidate stem cells were then cultured under conditions under which the generation of progenitors and mature cells could be quantitated. As in repopulation studies in vivo, functional capacity was inversely related to chronologic age of the donor (Lansdorp et al. 1993).

5.4
Genetic Influences

To complicate matters, it is now known that significant genetic differences exist between mouse strains that affect age-related changes in stem cell populations, making it difficult to compare results obtained by groups using different strains. For instance, we have compared C57BL/6 (B6) and DBA/2 (DBA) stem cell numbers and their cell cycle kinetics from fetal stages to 20-month-old adults (de Haan and Van Zant 1999a). Two significant changes were apparent. First, as has generally been found in other tissues during aging, the fraction of S-phase cells steadily decreased until at about 1 year it was below the sensitivity of the assay for both strains. At all earlier time points when it was measurable, as much as a ten-fold higher proportion of DBA cells was in S-phase. Second, stem cell numbers in both strains in-

Fig. 2. Longevity of mouse strains is inversely related to proliferation in a hematopoietic population. Variation in cell cycle kinetics of a progenitor cell population accounts for two-thirds of the longevity variation in eight commonly used inbred mouse strains. This relationship amongst recombinant inbred strains was used to map quantitative trait loci contributing to cycling and longevity, and to show that the same genomic intervals contributed to both traits

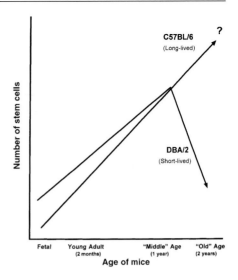

creased from fetal stages to young adulthood and to middle age (1 year) (Fig. 2). However, between 12 and 20 months a strain-specific difference was revealed. Whereas the number of B6 stem cells continued to increase at the same rate, the number of DBA stem cells declined by about half, to levels found in fetal liver. The relationship between these findings and the established longevity differences between these strains is taken up in a following section (7).

Stem cells rely heavily on interactions with stroma in vivo for a number of extrinsic signals favoring their survival and affecting both their proliferation and differentiation. Effects on stroma are therefore important in any considerations of age-related changes in stem cells. Although not as extensively studied as stem cells, there is evidence that cells of the stroma undergo a decline in proliferative capacity in aging that is functionally matched by a decline in their capacity to support of stem cells (Boggs et al. 1991).

6
Telomeres

6.1
Relationship to Replicative Senescence

An important experimental tool that has lent support to an intrinsic mechanism being related to replicative exhaustion of stem cells is the analysis of telomeres. The reader is referred to extensive reviews of this expanding field (Blackburn and Greider 1995; Chadwick and Cardew 1997). In brief, telomeres are the nucleotide repeats ($TTAGGG_n$ in human) that cap and stabilize the ends of chromosomes, preventing chromosomal fusions. They also solve the end-duplication problem of DNA synthesis identified by Watson and Olovnikov by providing the extension scaffold on which the DNA synthetic machinery can initiate complete replication of the genome (Olovnikov 1971; Watson 1972). In so doing, telomeres themselves shorten with each cell division, providing from their length a replicative history of the cell. The studies by Hayflick in the early 1960s showed that human fibroblasts have limited life spans of roughly 50 doublings in culture (Hayflick and Moorhead 1961), and it is now known that when telomeres shorten to a minimum critical length, cells become senescent, as did mitotically exhausted fibroblasts in Hayflick's cultures (Harley et al. 1990). A few cell types are able to synthesize telomerase, an enzyme which lengthens telomeres, but most normal somatic cells, including fibroblasts, are unable to do so (Liu et al. 1999). It has been convincingly shown that overexpression of telomerase in normal cultured cells allows them to escape replicative senescence apparently permanently, but without becoming tumorigenic (Bodnar et al. 1998). These findings firmly link preservation of telomeres with immortality, at least in vitro.

6.2
Telomeres and Stem Cells

There is ample evidence that hematopoietic stem cells produce telomerase, but apparently not enough to prevent telomere erosion after extensive proliferation (Morrison et al. 1996). Vaziri et al. (1994) showed that the length of telomeres of purified stem cells varied inversely with the age of the donor. Human stem cells from older adults had significantly shorter telomeres than their counterparts purified from umbilical cord blood, which, in turn, had telomeres shorter than from fetal liver stem cells.

6.2.1
Telomere Changes After Stem Cell Transplantation

In the transplantation setting, telomere length measurements have been made on blood cells of allografted patients and their stem cell donors. Telomeres were generally shorter in blood cells produced by transplanted stem cells than in blood cells originating from the same stem cell pool, but remaining in the donor (Notaro et al. 1997; Wynn et al. 1998). These data suggest that the telomere differences reflect the replicative stress undergone by the transplanted stem cells during engraftment. Moreover, the size of the telomere discrepancy was inversely proportional to the number of stem cells transplanted (Notaro et al. 1997). Thus the overall replicative stress on the population was influenced by the number of stem cells participating in engraftment. The fewer the stem cells in the graft, the greater the replicative stress on the stem cell population, and vice versa.

6.3
Telomeres and Aging

Although telomere length is now known to limit replicative potential of cells in vitro, the notion of its regulatory role in organismal longevity has been seductive but unproven in complex organisms such as mammals. That has now changed. Much of our recent knowledge concerning the physiological role of telomeres in vivo has come from mice null for telomerase. Possibly due to inherently long telomeres of mice, overt pathology was not detectable in the first six generations of telomerase null animals (Blasco et al. 1997). However, the sixth generation was unable to reproduce due to a lack of functional germ cells. Closer examination of this and earlier generations revealed subtle but significant changes (Rudolph et al. 1999). These were especially noteworthy in the continuously renewing tissues in which stem cells were the developmental cellular source of mature cells. In cells of several tissues, karyotypic analyses demonstrated potential sources of genetic instability in the form of extensive chromosomal abnormalities, including fusions.

Detailed longitudinal studies of telomerase knockout mice have now shown that particularly sixth-generation animals have shortened life spans and prematurely undergo physiological changes associated with aging in some but not all tissues (Rudolph et al. 1999). As in aged individuals, homeostatic functioning of most organ systems remains intact, but responses to physiologic stresses such as wound healing and hematopoietic recovery after chemotherapy is dramatically blunted. Thus, it appears that regenerative reserves in renewing tissues may be limited through effects on their respective stem cell populations.

6.4
Telomeres and Cancer

A striking finding in later generation (generations 3–6) telomerase knockout mice was the progressive increase in the rate of spontaneous tumors (Rudolph et al. 1999). Such increases were not only directly related to the generation but to the age of individuals within a generation. Telomeres normally serve to prevent chromosomal end-to-end fusions, and since such abnormalities occur at increased frequency in knockout mice, this serves as an attractive mechanism for their increased cancer incidence. It is also in accordance with the view that telomeres serve a tumor suppressor function both by preventing genetic instability and by imposing replicative limits on cells. In apparent contradiction is the observation that telomerase is usually upregulated in cancer cells (Shay 1997), providing a possible mechanism for their escape from replicative senescence, but apparently at cross purposes, also providing a measure of antitumor effect afforded by providing genetic stability. Carcinogenesis is obviously a complex process consisting of multiple steps which must be viewed in toto. Because telomerase is induced or upregulated in the process does not necessarily mean that it is the cause; it may as well be the effect of another step in the process. Mice doubly null for telomerase and the tumor suppressor INK4a$^{\Delta2/3}$ showed a decreased incidence of spontaneous tumor formation in vivo and decreased cellular oncogenic potential in vitro when compared to the mice lacking either gene alone (Greenberg et al. 1999). This effect was seen only in late-generation knockout mice when telomeres were already critically short. When telomerase activity was restored in cells from such animals, their oncogenic potential was also restored, indicating that telomerase induction was coincident with tumorigenesis in cells already possessing short telomeres. Thus, intact telomeres and telomerase synthetic pathway are cooperative events in carcinogenesis in this genetic setting and emphasize the paradoxical consequences of telomere regulation under different conditions and stages of cancer progression. They also demonstrate the potential pitfalls of focusing solely on, for example, telomerase inhibition as a cancer treatment.

7
A Link Between Stem Cell Replication and Organismal Life Span in the Mouse

7.1
Study of Embryo-Aggregated Chimeric Mice

The author's lab has been studying interstrain differences in the hematopoietic populations of mice, particularly as they relate to aging. These studies were prompted by the observation made with a long-time collaborator, Mike Dewey at the University of South Carolina, that early hematopoietic progenitor cells of different mouse strains varied widely with respect to cell cycle kinetics. The spleen colony-forming cell (CFU-S) populations of young adult mice B6 and DBA strains normally differed by ten-fold with respect to the population fraction in S-phase under steady-state conditions (Van Zant et al. 1983). Yet these strains have very similar blood cell counts. Chimeras (allophenic mice) were constructed by combining preimplantation-stage embryos of the two strains in order to determine if these and possibly other cryptic or subtle hematologic differences were cell-autonomous. Extrinsically regulated differences in stem cell function would be obviated in chimeras where stem cells of the two strains share a common environment. Our principal interest was in uncovering genetic differences manifested in the developmentally early cells themselves since they would certainly provide better insight into stem cell biology and may shed light on stem cell diseases, including leukemia. Initial studies of these chimeras revealed strain-specific susceptibility to leukemia induced by Friend murine leukemia virus (Eldridge and Dewey 1986), and a still unexplained skewing in uninfected animals of blood cell lineages (Van Zant et al. 1983). When normalized for overall chimerism in the hematopoietic system, DBA red cell and platelet formation was disproportionately high and neutrophil and especially lymphocyte production was skewed low. Long-term studies of chimerism in the blood formation were carried out because it was anticipated that subtle, cell-autonomous differences in the organization of the respective stem and progenitor populations might appear only after many months or in aged animals.

Another impetus for long-term studies of chimeras was the knowledge that DBA was a relatively short-lived strain and B6 was a long-lived strain. Invoking Hayflick's cell mitotic limit, we wondered if the rapidly proliferating DBA stem cell population might be replicatively exhausted in vivo and in some way contribute to the shorter life span of this strain. If the cell cycle phenotype was cell-autonomous it might be manifested in chimeras by a cessation of hematopoiesis derived from the DBA stem cells and blood formation would become wholly B6 in origin. This is in fact what was found in most chimeras (Van Zant et al. 1990). After about a year of stable blood cell chimerism, DBA representation in blood cells began to diminish and in many animals was completely absent after 2 years – approximately the normal the life span of this strain. Chimeras did not die at this time because the longer-lived B6 strain took over all blood cell production.

7.2
Reversibility of Stem Cell Activation and Quiescence

We next asked whether the DBA stem cells had disappeared completely,or whether some or all had entered a state of quiescence. To answer this, we harvested marrow from chimeras whose blood cells had become entirely B6 in origin, and transplanted it into irradiated recipients to determine the origin of engraftment. Surprisingly, cryptic DBA stem cells initially contributed heavily to engraftment but, after several months became completely quiescent again (Van Zant et al. 1992). Subsequent transplantation of marrow from primary to secondary recipients revealed the same pattern: DBA stem cells were reactivated and participated in blood formation, albeit to a lesser extent than in primary hosts. Again, after an even shorter period they became quiescent and all blood cells were of B6 origin. The fact that the limited extent of DBA chimerism and its short duration after secondary transplant argued for a DBA-specific decline in stem cell function. It was now clear that at least some stem cell phenotypes were cell – autonomous or intrinsic, thus opening the way for genetic studies.

7.3
Genetic Studies

7.3.1
In Vitro Assay for Stem Cells

A reliable and sensitive assay for stem cells was needed in order to embark upon genetic studies to map, clone and characterize genes responsible for the observed stem cell phenotypes. Since screening large numbers of mice would be required for the next step, an in vitro assay was chosen for practical reasons and cost considerations. The assay was based on the long-term bone marrow cell culture technique originally developed by Dexter and Lajtha (1974), and subsequently refined into a quantitative assay by Cashman et al. (1985), and Ploemacher et al. (1991). The cobblestone area forming cell (CAFC) assay takes advantage of the fact that stem cells require a close association with stromal cells in order to survive and function. To this end, limiting dilutions of a stem cell source are added to cultures of established stromal cell monolayers. We used a mouse line, FBMD-1, but primary stroma established from hematopoietic organs, or other cell lines work as well. In-dividual stem cells translocate beneath the stroma and after a period of quiescence, proliferate to form small colonies (5–50 cells) that appear as cobblestones under phase contrast illumination under the microscope. The length of the period of quiescence is variable for CAFCs and it has been shown by Ploemacher's group that quiescence is directly proportional to the developmental "primitiveness" of the stem cell (Ploemacher et al. 1991). In this way subpopulations of stem cells within a hierarchy can be delineated: the less primitive are more numerous, have a higher fraction in S-phase and form colonies in about a week. The most primitive have ~100-fold lower frequency, are largely out of cell cycle, and are quiescent for more than a month before forming colonies. Single cell sorting experiments have shown that CAFCs form colonies only once and at a time commensurate with their position in the hierarchy.

To obtain comparative interstrain data, we first measured the numbers of CAFCs on days 7, 14, 21, 28, and 35 in the bone marrow of eight commonly used inbred mouse strains. Whereas the numbers of day 7 cells was nearly uniform, numbers of day 35 cells varied by nearly ten-fold amongst mouse strains (de Haan and Van Zant 1997a,b). As we had seen before using the

CFU-S assay, the fraction of day 7 CAFCs in S-phase varied by more than tenfold amongst strains. Only AKR mice had a detectable number of the more primitive CAFC-day 35 in S-phase and, interestingly, these mice are very short-lived and die of lymphoma.

7.3.2
Average Life Span Correlates with Cell Cycle Kinetics

Surprisingly, when CAFC-day 7 cycling was plotted against the mean life spans of the eight strains, a highly significant inverse correlation was apparent such that nearly two-thirds of the variability in life span amongst strains was accounted for by differences in the cycling phenotype (Fig. 3; de Haan and Van Zant 1997a,b). These findings reinforced our hypothesis derived from the study of chimeras, that the two phenotypes were closely related, had a strong genetic component, and were amenable to genetic analysis.

B6 and DBA mice were chosen for further genetic analysis for several reasons. First, they had been the subject of our previous chimeric work. Second, using the CAFC assay the two strains consistently displayed nearly the largest differences in cycling, stem cell numbers, and life span. Third, powerful genetic resources were available to study these strains. Most significantly, a set of 26 (recently expanded to 42) recombinant inbred (BXD) mouse strains had been established at The Jackson Laboratory and were commercially available. As a result of genetic studies in these strains over the past 20 years by a large number of labs studying all manner of genetic traits, over 2000 loci and molecular markers have been mapped in these strains. Moreover, advanced intercrosses were ongoing between the two strains and were nearing the tenth generation (R. Williams, pers. comm.). The latter, because of the greatly increased genetic recombination, provide high-level detail for fine mapping.

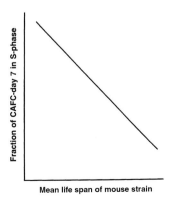

Fig. 3. "Old" age is relative to genetic background. In this example, stem cell number declines in the second year of life in short-lived DBA/2 mice whereas it continues to increase until at least 2 years in long-lived C57BL/6 mice. If declining stem cell number was causally related to longevity, a similar decline in C57BL/6 stem cells would be expected at about 3 years. The experiment is underway, but not completed

7.3.3
Mapping Studies in Recombinant Inbred Mouse Strains

Consequently, we phenotyped the BXD strains for cell cycling and CAFC numbers (de Haan and Van Zant 1997b). We were particularly interested in the relationship we had found previously between cycling and life span in the eight unrelated or distantly related inbred strains. To this end, we found that the difference in cycling between B6 and DBA was essentially accounted for by four quantitative trait loci (QTLs), a major locus on chromosome 11, and three minor loci on chromosomes 4, 7, and 9. Fortuitously, Gelman et al. (1988) had previously carried out longevity studies on the BXD strains more than 10 years ago. By plotting their longevity data against our cell cycling data, we found the same inverse correlation that we had found before; that is, the higher the cycling rate, the shorter the life span (Fig. 3). At the time the longevity studies were published, the genetic map was not sufficiently detailed nor were sophisticated mapping tools available for mapping QTLs.

7.3.4
Quantitative Trait Loci Affecting Life Span and Cell Cycle Kinetics Map to the Same Genomic Locations

We therefore remapped their life span data and found that four QTLs essentially accounted for the differences between B6 and DBA: two major loci on chromosomes 2 and 7, and two minor loci on chromosomes 4 and 11 (de Haan and Van Zant 1999b). The loci on chromosomes 7 and 11 mapped to the same locations as did the QTLs regulating CAFC cycling. Thus, two of the four QTLs contributing to each trait mapped to the same genomic intervals, suggesting, but not proving, a cause-and-effect relationship. Ongoing studies are aimed at generating strains congenic for each of the common intervals wherein the B6 segments are being moved onto a DBA genetic background and, reciprocally, the DBA segments are being moved onto a B6 background. The accuracy of the mapping coordinates of at least the primary QTL determining CAFC cycling is attested to by the phenotype of both the first and second backcross generations (de Haan et al. unpubl.). The presence of the appropriate DBA segment on chromosome 11 in a progressively more homozygous B6 genetic background is associated with increased cycling of the same magnitude as the difference between the two parental strains. We are using a 'speed' congenic strategy in which breeders for each generation are selected not only for the presence of the appropriate derivation of the interval of interest, but also for the least heterozygosity throughout the rest of the genome (Markel et al. 1997; Wakeland et al. 1997). This approach reduces to four or five the number of backcross generations required to reach a genetic background less than 1% heterozygous. Once congenics have been generated for each of the QTLs of interest, the phenotypic effects of each can be studied in isolation and can be selectively recombined by simply crossing congenic strains. Congenic mice will also facilitate the fine mapping and ultimate cloning of the QTLs.

7.3.5
Mapping of a Locus Determining Variation in Mouse Life Span

In the course of reanalyzing the longevity data from the BXD recombinant inbred strains, we noticed large disparities amongst the strains in the rate at which mice within a given strain died. The time from the death of the first mouse and the last mouse in a given strain was not correlated with their mean life spans, thus short-lived strains could have a large range, and vice versa. We therefore mapped range independently of mean life span and found that it was largely accounted for by a single genomic interval at the centromeric end of chromosome 11, a large distance from the QTL contributing to variation in mean life span between the two strains (de Haan et al. 1998). The former trait is novel in that it imparts variability in the phenotypic expression of a trait among genetically identical individuals of an inbred strain, a genetic concept that deserves further consideration as a general regulatory mechanism. Whether this locus, or others like it, affect expression of complex traits other than life span is not yet known. We are in the process of creating strains congenic for this interval, but are faced with a difficult phenotype to follow in genetic analyses.

8
Conclusions and Final Thoughts

Genetic determinants of longevity have proven surprisingly straightforward to unravel in yeast, flies, worms, and even in mammals. Using the limited context of life-span differences between selected mouse strains, a handful of QTLs account for at least the major differences. Given the advances in genetic analysis in the last few years and with whole genome sequences soon to be available for human and mouse, the prospects for cloning longevity affecting genes are promising. These heady advances in genetic analysis notwithstanding, it should be remembered that the only present manipulation that can increase longevity, albeit modestly, in any mammal is restricting caloric intake (Sprott 1997).

The finding that cell cycle kinetics of hematopoietic progenitor cells are related to organismal life span in mice is provocative since it provides at least a conceptual link between replicative senescence in vitro and organismal lifespan (Campisi 1996). It remains to be seen whether or not hematopoietic progenitor cell cycling is the sole population affected by the putative loci and therefore that the fate of these cells affects longevity, or if it is merely the only cell population analyzed to date. It is possible that the QTLs affecting cycling in this population have similar effects on other stem cell populations that may (too) be life span limiting. Since maintenance of telomere length in cultured cells delays or prevents replicative senescence, it is reasonable to suggest that telomere shortening may be involved in the decline of certain tissues in vivo as well. In support of this notion, studies of telomerase knockout mice have shown that it is the renewing adult tissues – presumably their respective stem cells – that are most affected. Whether or not QTLs affecting cell cycling and life span specify some aspect of telomere regulation remains to be seen.

It is fitting that stem cells of a continuously renewing tissue would be implicated in organismal longevity. Development, aging and longevity are all facets of the same natural state: life. The keys to better understanding longevity and aging will most likely be found in the study of development.

Acknowledgments. The studies from the author's lab were partially supported by NIH grant R01 AG16653. Gerald de Haan is a fellow of The Netherlands Organization for Scientific Research and of the Royal Dutch Academy of Arts and Sciences.

References

Abkowitz JL, Lineberger ML, Newton MA, Shelton GH, Ott RL, Guttorp P (1990) Evidence for the maintenance of hematopoiesis in a large animal by the sequential activation of stem-cell clones. Proc Natl Acad Sci USA 87:9062–9066

Abkowitz JL, Taboada M, Shelton GH, Catlin SN, Guttorp P, Kiklevich JV (1998) An X chromosome gene regulates hematopoietic stem cell kinetics. Proc Natl Acad Sci USA 95:3862–3866

Anderson JE, Gooley TA, Schoch G, Anasetti C, Bensinger WI, Clift RA, Hansen JA, Sanders JE, Storb R, Appelbaum FR (1997) Stem cell transplantation for secondary acute myeloid leukemia: evaluation of transplantation as initial therapy or following induction chemotherapy. Blood 89:2578–2585

Beddington RSP, Robertson EJ (1999) Axis development and early asymmetry in mammals. Cell 96:195–209

Bjornson CRR, Rietze RL, Reynolds BA, Magli MC, Vescovi AL (1999) Turning brain into blood: A hematopoietic fate adopted by adult neural stem cells in vivo. Science 283:534–537

Blackburn EM, Greider CW (eds) (1995) Telomeres. Cold Spring Harbor Laboratory Press, Cold Spring Harbor, New York

Blasco MA, Lee HW, Hande MP, Samper E, Lansdorp PM, DePinho RA, Greider CW (1997) Telomere shortening and tumor formation by mouse cells lacking telomerase RNA. Cell 91:25–34

Bodnar G, Ouellette M, Frolkis M, Holt SE, Chiu C-P, Morin GB, Harley CB, Shay JW, Lichtsteiner S, Wright WE (1998) Extension of life-span by introduction of telomerase into normal human cells. Science 279:349–352

Boggs SS, Patrene KD, Austin CA, Vecchini F, Tollerud DJ (1991) Latent deficiency of the hematopoietic microenvironment of aged mice as revealed in W/W^v mice given +/+ cells. Exp Hematol 19:683–687

Bohr V, Anson RM, Mazur S, Dianov G (1998) Oxidative DNA damage processing and changes with aging. Toxicol Lett 103:47–52

Bradford GB, Williams B, Rossi R, Bertoncello I (1997) Quiescence, cycling, and turnover in the primitive hematopoietic stem cell compartment. Exp Hematol 25:445–453

Buller RE, Sood AK, Lallas T, Buekers T, Skilling JS (1999) Association between nonrandom X-chromsome inactivation and BRCA1 mutation in germline DNA of patients with ovarian cancer. J Natl Cancer Inst 91:339–346

Campisi J (1996) Replicative senescence: an old lives' tale? Cell 84:497–500

Capel B, Hawley R, Covarrubias L, Hawley T, Mintz B (1989) Clonal contributions of small numbers of retrovirally marked hematopoietic stem cells engrafted in unirradiated neonatal W/Wv mice. Proc Natl Acad Sci USA 86:4564–4568

Capel B, Hawley RG, Mintz B (1990) Long- and short-lived murine hematopoietic stem cell clones individually identified with retroviral integration markers. Blood 75:2267–2270

Cashman J, Eaves AC, Eaves CJ (1985) Regulated proliferation of primitive hematopoietic progenitor cells in long-term human marrow cultures. Blood 66:1002–1005

Chadwick DJ, Cardew G (eds) (1997) Telomeres and telomerase. John Wiley and Sons, New York

Chen J, Astle BA, Harrison DE (1999) Development and aging of primitive hematopoietic stem cells in BALB/cBy mice. Exp Hematol 27:928–935

Cheshier SH, Morrison J, Liao X, Weissman IL (1999) In vivo proliferation and cell cycle kinetics of long-term self-renewing hematopoietic stem cells. Proc Natl Acad Sci USA 96:3120–3125

Clark MR (1988) Senescence of red blood cells: progress and problems. Physiol Rev 68:503–554

de Haan G, Nijhof W, Van Zant G (1997a) Mouse strain-dependent changes in frequency and proliferation of hematopoietic stem cells during aging: Correlation between lifespan and cycling activity. Blood 89:1543–1550

de Haan G, Van Zant G (1997b) Intrinsic and extrinsic control of hemopoietic stem cell numbers: Mapping of a stem cell gene. J Exp Med 186:529–536

de Haan G, Gelman R, Watson A, Yunis E, Van Zant G (1998) A putative gene causes variability in lifespan among genotypically identical mice. Nat Genet 19:114–116

de Haan G, Van Zant G (1999a) Dynamic changes in mouse hematopoietic stem cell numbers during aging. Blood 93:3294–3301

de Haan G, Van Zant G (1999b) Genetic analysis of hemopoietic cell cycling in mice suggests its involvement in organismal life span. FASEB J 13:707–713

Dexter TM, Lajtha LG (1974) Proliferation of haemopoietic stem cells in vitro. Br J Haematol 28:525–530

Dick JE, Lapidot T, Vormoor J, Larochelle A, Bonnet D, Wang J (1995) Human hematopoiesis in SCID mice. In: Gluckman E, Coulombel L (eds) Ontogeny of hematopoiesis. E. Gluckman and L. Coulombel, eds. INSERM, Vandoeuvre-les-Nancy, pp. 97–101

Effros RB (1998) Replicative senescence in the immune system: impact of the Hayflick limit on T-cell function in the elderly. Am J Hum Genet 62:1003–1007

Eldridge PW, Dewey MJ (1986) Genotype-limited changes in platelet and erythroid kinetics in Friend-virus-infected allophenic mice. Exp Hematol 14:380–385

Fraser CC, Eaves CJ, Szilvassy SJ, Humphries KR (1990) Expansion of retrovirally marked totipotent hematopoietic stem cells. Blood 76:1071–1076

Gale E, Fielding K, Harrison CN, Linch DC (1997) Acquired skewing of X-chromosome inactivation patterns in myeloid cells of the elderly suggests stochastic clonal loss with age. Br J Haematol 98:512–519

Geiger H, Sick S, Bonifer C, Muller AM (1998) Globin gene expression is reprogrammed in chimeras generated by injecting adult hematopoietic stem cells into mouse blastocysts. Cell 93:1055–1065

Gelman R, Watson, Bronson R, Yunis E (1988) Murine chromosomal regions correlated with longevity. Genetics 118:693–704

Globerson A (1999) Hematopoietic stem cells and aging. Exp Gerontol 34:137–146

Goldwasser E (1975) Erythropoietin and the differentiation of red blood cells. Fed Proc 34:2285–2292

Greenberg RA, Chin L, Femino A, Lee K-H, Gottlieb GJ, Singer RH, Greider CW, DePinho RA (1999) Short dysfunctional telomeres impair tumorigenesis in the INK4a$^{\Delta2/3}$ cancer-prone mouse. Cell 97:515–525

Gurdon JB (1968) Transplanted nuclei and cell differentiation. Sci Am 219:24–35

Harley CB, Futcher AB, Greider CW (1990) Telomeres shorten during ageing of human fibroblasts. Nature 345:458–460

Harrison E (1972) Normal function of transplanted mouse erythrocyte precursors for 21 months beyond donor life spans. Nat New Biol 237:220–222

Harrison E (1980) Competitive repopulation: a new assay for long-term stem cell functional capacity. Blood 55:77–81

Harrison DE (1983) Long-term erythropoietic repopulating ability of old, young and fetal stem cells. J Exp Med 157:1496–1504

Harrison DE, Astle CM, Delaittre JA (1978) Loss of proliferative capacity in immunohemo-poietic stem cells is caused by serial transplantation rather than aging. J Exp Med 147:1526–1531

Harrison DE, Lerner C, Hoppe PC, Carlson GA, Alling D (1987) Large numbers of primitive cells are active simultaneously in aggregated embryo chimeric mice. Blood 69:773–777

Harrison E, Astle CM, Stone M (1989) Effects of age on transplantable primitive immuno-hematopoietic stem cell (PSC) numbers and function. J Immunol 142:3833–3840

Hayflick L (1998) How and why we age. Exp Gerontol 33:639–653

Hayflick L, Moorhead PS (1961) The serial cultivation of human diploid cell strains. J Exp Cell Res 25:585–621

Hsin H, Kenyon C (1999) Signals from the reproductive system regulate the lifespan of C. elegans. Nature 399:362–366

Iscove NN, Nawa K (1997) Hematopoietic stem cells expand during serial transplantation in vivo without apparent exhaustion. Curr Biol 7:805–808

Johansson CB, Momma S, Lendahl U, Frisen J (1999) Identification of a neural stem cell in the adult mammalian central nervous system. Cell 96:25–34

Johnson FB, Sinclair DA, Guarente L (1999) Molecular biology of aging. Cell 96:291–302

Jordan CT, Lemischka IR (1990) Clonal and systemic analysis of long-term hematopoiesis in the mouse. Genes Dev 4:220–232

Kay HEM (1965) How many cell generations? Lancet ii: 418

Keller G, Snodgrass R (1990) Life span of multipotential hematopoietic stem cells in vivo. J Exp Med 171:1407–1418

Koury MJ (1992) Programmed cell death (apoptosis) in hematopoiesis. Exp Hematol 20:391–394

Lansdorp PM, Dragowska W, Manyani H (1993) Ontogeny-related changes in proliferative potential of human hematopoietic cells. J Exp Med 178:787–791

Liu K, Schoonmaker MM, Levine BL, June CH, Hodes RJ, Weng N (1999) Constitutive and regulated expression of telomerase reverse transcriptase (hTERT) in human lymphocytes. Proc Natl Acad Sci USA 96:5147–5152

Loeffler M, Potten CS (1997) Stem cells and cellular pedigrees – a conceptual introduction. In: Potten CS (ed) Stem cells. C.S. Potten, ed. Academic Press, London, pp. 1–27

Mannervik M, Nibu Y, Zhang H, Levine M (1999) Transcriptional coregulators in development. Science 284:606–609

Markel P, Shu P, Ebeling C, Carlson GA, Nagle DL, Smutko JS, Moore KJ (1997) Theoretical and empirical issues for marker-assisted breeding of congenic mouse strains. Nat Genet 17: 280–284

Martin G, Austad SN, Johnson TE (1996) Genetic analysis of ageing: role of oxidative damage and environmental stresses. Nat Genet 13:25–34

Mauch P, Botnick LE, Hannon EC, Obbagy J, Hellman S (1982) Decline in bone marrow proliferative capacity as a function of age. Blood 60:245–252

Metcalf D (1988) The molecular control of blood cells. Harvard University Press, Cambridge, Massachusetts

Micklem HS, Ford CE, Evans EP, Ogden DA, Papworth DS (1972) Competitive in vivo proliferation of foetal and adult hematopoietic cells in lethally irradiated mice. J Cell Physiol 79:293–298

Micklem HS, Lennon JE, Ansell JD, Gray RA (1987) Numbers and dispersion of repopulating hematopoietic cell clones in radiation chimeras as functions of injected cell dose. Exp Hematol 15:251–257

Miller RA (1996) The aging immune system: primer and prospectus. Science 273:70–74

Morrison SJ, Prowse KR, Ho P, Weissman IL (1996a) Telomerase activity in hematopoietic cells is associated with self-renewal potential. Immunity 5:207–216

Morrison SJ, Wandycz AM, Akashi K, Globerson A, Weissman IL (1996b) The aging of hematopoietic stem cells. Nat Med 2:1011–1016

Morrison SJ, White PM, Zock C, Anderson DJ (1999) Prospective identification, isolation by flow cytometry, and in vivo self-renewal of multipotent mammalian neural crest stem cells. Cell 96:737–749

Notaro R, Cimmino A, Tabarini D, Rotoli B, Luzzatto L (1997) In vivo telomere dynamics of human hematopoietic stem cells. Proc Natl Acad Sci USA 94:13782–13785

Ogden DA, Micklem HS (1976) The fate of serially transplanted bone marrow cell populations from young and old donors. Transplantation 22:287–293

Olovnikov AM (1971) Principles of marginotomy in template synthesis of polynucleotides. Dokl Akad Nauk SSSR 201:1496–1499

Osawa M, Hanada K-I, Hamada H, Nakauchi H (1996) Long-term lymphohematopoietic reconstitution by a single CD34-low/negative hematopoietic stem cell. Science 273:242–245

Osgood EE (1957) A unifying concept of the itiology of the leukemias, lymphomas, and cancers. J Natl Cancer Inst 18:155–166

Peterson BE, Bowen WC, Patrene KD, Mars WM, Sullivan AK, Murase N, Boggs SS, Greenberger JS, Goff JP (1999) Bone marrow as a potential source of hepatic oval cells. Science 284:168–170

Pietrzyk ME, Priestley GV, Wolf NS (1985) Normal cycling patterns of hematopoietic stem cell populations: an assay using long-term in vivo BrdU infusion. Blood 66:1460–1462

Ploemacher RE, Brons NHC (1988) Isolation of hemopoietic stem cell subsets from murine bone marrow: I. Radioprotective ability of purified cell suspensions differing in the proportion of day-7 and day-12 CFU-S. Exp Hematol 16:21–26

Ploemacher RE, van der Sluijs JP, van Beurden CA, Baert MR, Chan PL (1991) Use of limiting-dilution type long-term marrow cultures in frequency analysis of marrow-repopulating and spleen colony-forming hematopoietic stem cells in the mouse. Blood 78:2527–2533

Potten CS (1997) Stem cells, Academic Press, London

Potten CS, Loeffler M (1990) Stem cells: attributes, cycles, spirals, pitfalls and uncertainties. Lessons for and from the crypt. Development 110:1001–1020

Rebel VI, Miller CL, Eaves CJ, Lansdorp PM (1996) The repopulation potential of fetal liver hematopoietic stem cells in mice exceeds that of their adult bone marrow counterparts. Blood 87:3500–3507

Ross E, Anderson N, Micklem HS (1982) Serial depletion and regeneration of the murine hematopoietic system. Implications for hematopoietic organization and the study of cellular aging. J Exp Med 155:432–444

Rossant J, Nagy A (1999) In search of the *tabula rosa* of human cells. Nat Biotechnol 17:23–24

Rubin H (1997) Cell aging in vivo and in vitro. Mech Ageing Dev 98:1–35

Rudolph KL, Chang S, Lee H-W, Blasco M, Gottlieb GJ, Greider CW, DePinho RA (1999) Longevity, stress response, and cancer in aging telomerase-deficient mice. Cell 96:701–712

Sattler M, Winkler T, Verma S, Byrne CH, Shrikhande G, Salgia R, Griffin JD (1999) Hematopoietic growth factors signal through the formation of reactive oxygen species. Blood 93:2928–2935

Schlessinger D, Ko MSH (1998) Developmental genomics and its relation to aging. Genomics 52:113–118

Shamblott MJ, Axelman J, Wang S, Bugg EM, Littlefield JW, Donovan PJ, Blumenthal PD, Huggins GR, Gearhart JD (1998) Derivation of pluripotent stem cells from cultured human primordial germ cells. Proc Natl Acad Sci USA 95:13726–13731

Shay JW (1997) Telomerase in human development and cancer. J Cell Physiol 173:266–270

Shiels PG, Kind AJ, Campbell KHS, Waddington D, Wilmut I, Colman A, Schnieke AE (1999) Analysis of telomere lengths in cloned sheep. Nature 399:316–317

Siminovitch L, Till JE, McCulloch EA (1964) Decline in colony-forming ability of marrow cells subjected to serial transplantation into irradiated mice. J Cell Comp Physiol 64:23–31

Smith LG, Weissman IL, Heimfeld S (1991) Clonal analysis of hematopoietic stem-cell differentiation in vivo. Proc Natl Acad Sci USA 88:2788–2792

Socie G (1996) Secondary malignancies. Curr Opin Hematol 6:466–70

Solter D, Gearhart J (1999) Biomedicine – Putting stem cells to work. Science 283:1468–1470

Spangrude GJ, Heimfeld S, Weissman IL (1988) Purification and characterization of mouse hematopoietic stem cells. Science 241:58–62

Spangrude GJ, Brooks DM, Tumas DB (1995) Long-term repopulation of irradiated mice with limiting numbers of purified hematopoietic stem cells: in vivo expansion of stem cell phenotype but not function. Blood 85:1006–1016

Sprott RL (1997) Diet and caloric restriction. Exp Gerontol 32:205–214

Szilvassy SJ, Fraser CC, Eaves CJ, Lansdorp PM, Eaves AC, Humphries RK (1989) Retrovirus-mediated gene transfer to purified hemopoietic stem cells with long-term lympho-myelo-poietic repopulating ability. Proc Natl Acad Sci USA 86:8798–8802

Thomson JA, Itskovitz-Eldor J, Shapiro SS, Waknitz MA, Swiergiel JJ, Marshall VS, Jones JM (1998) Embryonic stem cell lines derived from human blastocysts. Science 282:1145–1147

Till JE, McCulloch EA, Siminovitch L (1964) A stochastic model of stem cell proliferation, based on the growth of spleen colony-forming cells. Proc Natl Acad Sci USA 51:29–36

Uchida N, Weissman IL (1992) Searching for hematopoietic stem cells: Evidence that Thy-1.1^{lo} Lin$^-$ Sca-1$^+$ cells are the only stem cells n C57BL/Ka-Thy-1.1 bone marrow. J Exp Med 175:175–184

Van Zant G, Eldridge PW, Behringer RR, Dewey MJ (1983) Genetic control of hematopoietic kinetics revealed by analyses of allophenic mice and stem cell suicide. Cell 35:639–645

Van Zant G, Holland BP, Eldridge PW, Chen J-J (1990) Genotype-restricted growth and aging patterns in hematopoietic stem cell populations of allophenic mice. J Exp Med 171:1547–1565

Van Zant G, Scott-Micus K, Thompson BP, Fleischman RA, Perkins S (1992) Stem cell quiescence/activation is reversible by serial transplantation and is independent of stromal cell genotype in mouse aggregation chimeras. Exp Hematol 20:470–475

Vaux DL, Korsmeyer SJ (1999) Cell death in development. Cell 96:245–254

Vaziri H, Dragowska W, Allsopp RC, Thomas TE, Harley CB, Lansdorp PM (1994) Evidence for a mitotic clock in human hematopoietic stem cells: loss of telomeric DNA with age. Proc Natl Acad Sci USA 91:9857–9860

Visser JWM, Bauman JGJ, Mulder AH, Eliason JF, de Leeuw AM (1984) Isolation of murine pluripotent hemopoietic stem cells. J Exp Med 159:1576–1590

Wakayama T, Yanagimachi R (1999) Cloning of male mice from adult tail-tip cells. Nat Genet 22:127–128

Wakayama T, Perry A, Zuccotti M, Johnson KR, Yanagimachi R (1998) Full-term development of mice from enucleated oocytes injected with cumulus cell nuclei. Nature 394:369–374

Wakeland E, Morel L, Achey K, Yui M, Longmate J (1997) Speed congenics: a classic technique in the fast lane (relatively speaking). Immunol Today 18:472–477

Warner CM, McIvor JL, Stephens TJ (1977) Chimeric drift in allophenic mice. Differentiation 9:11–17

Watson JD (1972) Origin of concatameric T7 DNA. Nat New Biol 239:197–201

Weismann A (1893) The germ-plasm: a theory of heredity. Walter Scott, London

Williams LH, Udupa KB, Lipschitz DA (1986) Evaluation of the effect of age on hematopoiesis in the C57BL/6 mouse. Exp Hematol 14:827–832

Wilmut I, Schnieke AE, McWhir J, Kind AJ, Campbell KHS (1997) Viable offspring derived from fetal and adult mammalian cells. Nature 385:810–813

Winton DJ, Ponder BA (1990) Stem-cell organization in mouse small intestine. Proc R Soc Lond B Biol Sci 241:13–18

Wynn RF, Cross MA, Hatton C, Will AM, Lashford LS, Dexter TM, Testa NG (1998) Accelerated telomere shortening in young recipients of allogeneic bone-marrow transplants. Lancet 351:178–181

Yu H, Bauer B, Lipke GK, Phillips RL, Van Zant G (1993) Apoptosis and hematopoiesis in murine fetal liver. Blood 81:373–384

Subject Index

Printing: Saladruck, Berlin
Binding: H. Stürtz AG, Würzburg